RADS, ERGS, AND CHEESEBURGERS

The Kids' Guide to Energy and the Environment

Bill Yanda

John Muir Publications
Santa Fe, New Mexico

John Muir Publications, P.O. Box 613, Santa Fe, NM 87504

First edition. First printing

Library of Congress Cataloging-in-Publication Data

Yanda, Bill.
   Rads, ergs, and cheeseburgers :   the kid's guide to energy and the
environment / Bill Yanda. — Ist ed.
      p. cm.
   Includes index.
   Summary: Ergon, a magical being, discusses the generation of
various forms of energy, its transportation, uses in everyday life,
conservation, and development of alternative sources.
   ISBN 0-945465-75-0
   1. Energy development—Environmental aspects—Juvenile literature.
2. Energy conservation—Juvenile literature.   3. Renewable energy
sources—Juvenile literature.   [1. Power resources.   2. Energy
conservation.] I. Title.
TD195.E49Y36   1991
333.79—dc20

                                                            90-22672
                                                               CIP
                                                                AC

Designer: Sally Blakemore
Illustrations: Michael Taylor
Typeface: Gill Sans and Maximal Display
Typesetter: Copygraphics, Santa Fe, New Mexico
Printer: Guynes Printing, Albuquerque, New Mexico

Printed on Recycled Paper

Distributed to the book trade by
W. W. Norton and Company, Inc.
New York, New York

There is a real life Katy and a real life Mark. Just like the ones in this book, they're about two years apart, with Katy being the older. Unlike the kids in this book, my Katy and Mark have grown up in a home that is entirely heated by a solar greenhouse and they understand a great deal about renewable energy. There is also a Rachel, who is ten years younger; her namesake doesn't show up anywhere in this book.

They are all great kids and I love them. May they, and you, grow up in a world like Verdant and not like one of the others.

And to Susan, who has been my wife, friend, business partner, and lifemate for twenty-three years.

And to the solar, conservation, and alternative energy experts and enthusiasts who haven't given up during the last ten years. Thank God there are some people who can see farther than the politicians.

CONTENTS

1

A RUDE SURPRISE

It was a bitter January evening. A strong cold front was blowing in from the north, and the winds drove dry leaves and twigs against the windows with an angry tapping. Katy and Mark had just gotten home from school. Their mom and dad wouldn't be home from work for two more hours, so the children were involved in their regular after-school routine. Mark had turned on the TV. Katy had started a fire in the fireplace. Mark also had his stereo going full blast in his room. Both kids wanted a bright house when they were by themselves, so every light had been turned on.

"I don't think it's warm enough in here," Katy mumbled, as she walked to the hall and turned the thermostat up to 85 degrees.

"Did you put my pizza in the toaster-oven?" Mark shouted to her.

"No, I'm heating a bagel in there," Katy answered. "I stuck your pizza in the oven and set it on high."

Mark was irritated that he would have to wait even longer for a snack. "Why didn't you put it in the microwave?"

"Because I have some frozen fruit salad thawing in there, and besides your pizza would be limp as spaghetti if we nuked it, frog face," Katy replied, flopping down in her dad's favorite chair.

"Don't call me frog face, mouse lips!" And with that Mark sent a big sofa cushion flying toward Katy's head.

Katy recovered from the sneak attack and leaped off the chair to bury Mark in a shower of smaller pillows and newspapers. Just as they were about to fall off the couch, Mark grasping Katy in a headlock, the wind hit the house like a truck, the walls shuddered, and the lights went out. For a moment the room went black, and both children clung to each other in silence. Only a small pale light from the fireplace in another room crept into the darkness.

The TV, stereo, and lights came back on in an instant. The kids untangled from each other with embarrassed quickness.

"I wasn't scared, but you were terrified," Mark teased, crawling back onto the sofa.

Katy was about to answer when her attention was drawn to the television.

"This is an NBC News Bulletin," the announcer said. "It has just been learned that an OMNI petroleum tanker has run aground off the coast of Maine. The circumstances seem similar to the huge oil spill in Alaska several years ago. Much of the coastline adjoining the wreckage site is National Wilderness Area, and damage to wildlife is expected to be widespread."

At this point the broadcast showed still photographs of oil-covered birds, seals, and fish in a pitiful condition. Katy buried her head in her arm at the site and Mark just said, "Eckggg."

"Coast Guard rescue vessels are rushing to the site, but it is believed that all of the crew got off the boat safely. Stay tuned to this station for further details as they become available." The program went back to some vehicle chase with cars and trucks flipping and rolling all over the highway.

"It just makes me sick," Katy groaned. "All those dead animals, for what? So some oil company can make more money."

"Yeah," Mark said with bitterness. "Why don't *they* just stop doing that? If *they* wanted to, *they* could. . ."

Suddenly, the house went black again. This time, there was no sound. Even the howling wind had disappeared.

"Katy, are you OK?" Mark whispered, groping around in the blackness to touch his sister. She wasn't within his reach.

"I. . .I'm f. . .f. . .fine," she stuttered, "but why's everything so quiet and so dark?"

As the word "dark" was leaving her lips, the TV screen began to glow. First, an eerie pale yellow. Then it changed to a yellow-orange shade and then to a deep orange, growing ever brighter and angrier-looking.

The children, easily visible to each other by the strange brightness of the TV screen, began to push themselves backward away from the threatening device.

"I think it's going to explode!" Mark called. His back was hard against the couch, and he could move no farther in that direction. His eyes were fixed on the screen, and, as he watched, a form was growing out of the brilliant orange glow, expanding and rising above the TV set.

"Protect your face, Mark," Katy cried, but the boy couldn't take his eyes from the hypnotic sight.

Now the thing, a pulsing orange mass of energy, was assuming a form. It looked to Katy (who couldn't really hide her face, either) as if large eyes were developing toward the top of the blob.

At this moment, all noise stopped again. The shape had disconnected itself from the TV screen and was in the process of growing human-looking legs and arms! It was tall, as tall as the ceiling and looking down on them.

"What is it?" Mark muttered under his breath.

Katy said nothing. Both sat perfectly still, backs rigid and stiff, as the being stared down at them. It seemed to the kids that a year went by before the creature finally spoke.

"I AM ERGON."

The voice was deep and rumbling, like the noise a river makes as it drops over a large waterfall. Neither Katy nor Mark could say a word.

"I AM ERGON. I AM PURE ENERGY. YOU ARE ENERGY ABUSERS, AND YOU DON'T EVEN KNOW WHAT YOU ARE DOING. THAT'S WHY I'M HERE." The being paused for a few moments.

"I SPEAK IN CAPITAL LETTERS WHEN I'M DISTURBED."

Mark squirmed in his seat and finally got up the courage to say softly, "I'm sorry, but we don't know what you're talking about."

"How are we energy abusers? We're just kids," Katy said quietly.

"YOU'RE OLD ENOUGH AND SMART ENOUGH TO USE ENERGY WISELY!" Ergon roared, causing both Katy and Mark to slink down and try to pretend they were invisible. There was a long pause as Ergon continued to look down on them with a disturbing glare.

"THINK about this." The children were relieved as the voice became quieter.

"You've turned on every light in this house." The room suddenly became bright once again as the lights all over the house came on.

"You've got five different heat sources going." The fire could be heard sparking back to life, the toaster-oven and microwave began purring, and the home heating system started up with a whoosh.

"And your **NOISE MACHINES** are adding to the energy drain."
The TV and Mark's stereo roared back their reply.

"AND YOU HAVE THE NERVE TO SAY OTHERS ARE THE ONLY ONES RESPONSIBLE FOR THE ACCIDENTS!"

Katy bravely answered, "We don't make oil tankers run aground or power plants explode."

"No. Energy companies and governments have not been concerned enough about environmental safety, that's true. But every time you use energy unwisely, you are adding directly to the problem." Ergon paused, as if considering something. **"I believe you two need to find out more about energy—how it's made, how it's moved, and how it's used. Yes, that's exactly what you need."**

Katy turned toward Mark and saw him shrinking and dissolving. At the same time, she felt her body begin to lose its "glue." It was as if her feet could be attached to her arms, or her head was stuck on just below her left kneecap. The familiar room had lost organization. The TV was falling into the couch, and dad's favorite chair was melting into the carpet.

She was drifting toward a shimmering darkness. Every part of her being was cold.

2

BARREN AND NOXIOUS

Katy began to feel reassembled at the outer arm of a spiral galaxy as she dashed between a red gas giant and nasty-looking black hole. She knew she wasn't a human anymore but a small and shining point of light. Mark and Ergon looked the same, tiny points of light shooting through a field of stars. Katy had seen lots of space movies before, but this was different. The colors, the depth, the sheer size and power of stars made her gasp.

"Where are you taking us? Why aren't we dead out here in space?" Mark wanted to know. His words weren't actually spoken but came out clearly to all of them.

"I've changed you both to energy like me," Ergon replied. **"That's why you aren't dead. I want you to see worlds where beings have used their planet's natural resources in different ways."**

Far ahead the children could see a small tan-colored ball growing larger. As they continued their headlong plunge, the world seemed to be entirely a desert, so bright it hurt the eyes to look directly on it. Ergon led the trio on a steep dive to the surface.

"This planet was once called Plenty. Now the few creatures left call it Barren," Ergon said.

"How could anyone live here?" Katy asked in wonder. "There are no clouds, no trees, no water. Nothing but sand and dust. Is this the way it's always been?"

In an instant the view changed and the kids were flying over a deep hole in the ground with slices cut along the walls and into the cavity. Behind the huge pit was an ugly mountain, scarred deeply along its sides with cuts and gouges.

"Those who lived here never learned that when you use raw material for energy, you must replant the trees you've cut, restore the earth, and always treat the planet like you'd treat your own backyard." Ergon landed the group in the lowest level of the pit, an area that looked like the inside of a bomb site. Ergon watched Katy and Mark as they looked around in dismay.

"But why did they do this?" Mark questioned.

"The beings on Barren discovered that its great natural resources —trees, minerals, oil, coal—could bring them lots of jobs and money. They sold all these resources to other planets. For a while, everyone's life was better as the entire world had jobs digging, cutting, and mining the land. As the resources got harder to find, they'd dig farther and cut faster. The beings didn't realize that topsoil, the thin layer of organic matter covering the planet, is the real wealth of a world. The top few inches of rich earth produce everything necessary for life on the surface.

"Because grasses and trees weren't replanted, rain eroded the land and carried all the topsoil down to the seas. In addition, the seas and lakes were getting the leftover chemicals from industrial production. Plants and animals of the oceans were slowly poisoned."

Katy and Mark looked at the great piles of rock and eroded earth around them. Mark had an idea.

"I'll bet the next thing that happened is that the desert started taking over the world," he said.

"That's correct. With no topsoil to hold the rain and grow new trees and grasses, more and more of the world became desert. In time, the entire world was barren."

"Why didn't they see what was happening? Maybe they could have stopped it in time?"

"It didn't happen all at once." Ergon explained that the slow erosion of the planet's fertile lands which brought death to the trees and grasses took hundreds of years. The desert grew only a few miles a year. The beings continued to cut down their forests and use up the good soil in farming and mining until it was too late to reverse the process.

Katy and Mark were looking around the blasted pit when they noticed small dark figures moving quickly along the edge high up near a cave. They squinted into the bright sunlight and noticed that the movement seemed to be down the sides of the hole and toward them. They could almost see the darting creatures, then all movement would stop, as the shapes disappeared into a dark crack or behind a rock. The scurrying forms were like cockroaches running across a kitchen counter looking for scraps to eat.

Katy turned to ask Ergon what they were, but Ergon was gone. Both kids moved a step closer to each other but didn't take their eyes off the dark shapes now moving lower and lower down the walls. They watched in silence and growing fear until Mark said what they both were feeling.

"He wouldn't leave us here," Mark whispered. "Or would he?"

Ergon returned in a crash of exploding air.

"I don't intend to. Look at this."

Katy and Mark forgot their fear and moved closer to Ergon. He had made one of his hands into a flat dishlike platform. On the shimmering surface was a small green leaf.

"Everybody's seen a leaf before," Mark said.

"Perhaps you've never thought about how important this is to your world. Take a closer look."

Suddenly, both children felt themselves shrinking and being drawn toward the leaf.

"Whoooa!" Mark yelled. "I can see just fine from where I am!"

In an instant, they were standing on a green plain that stretched out in all directions as far as they could see. Below their feet was a complex pattern of regular shapes that dissolved into the distance. The surface seemed to be pulsating with a powerful rhythm, as if a hundred jets were taking off at once. Above that noise, they heard Ergon's voice, far away but clear.

"You are standing on the most important energy machine ever created."

"But it's just a leaf," Mark replied. "A big one, but still a leaf. On earth we've got power plants and refineries that are huge. You could put a million leaves into. . ."

"WITHOUT THIS LEAF, YOU WOULDN'T HAVE ANY POWER PLANTS, OR ANY POWER AT ALL." Ergon's voice boomed, causing the surface of the leaf to roll and both children to fall on their behinds.

"The leaf, and its cousins, grass, sea algae, and floating sea plants, convert light from the sun into different forms of energy."

As Katy and Mark listened, Ergon explained how at the surface of green plants, sunlight, water, and carbon dioxide (CO_2) are changed into chemical energy. This process is called *photosynthesis*. It converts energy in sunlight into carbon-based molecules, or carbohydrates. The carbohydrates are the basis for food as well as wood, coal, and oil. During the process, the leaf absorbs CO_2 from the air and gives back oxygen, which animals and people need to live. When the process is reversed, it is called *respiration*, the result being the release of water, oxygen, and heat. When the plant dies, its body enriches the soil or ocean, making way for more plants and animals to live.

"Oil, gas, coal—all of the major forms of energy are made from plants. On earth, you're living off the stored energy living things made millions of years ago. That's why these energy sources are called fossil fuels. They are nonrenewable, which means they can't be replaced in the future."

Just as quickly as they'd been transported to the leaf, Ergon enlarged the children, and they found themselves back in the bright sunlight of the pit. Katy and Mark looked around them and saw that the dark shapes were closer now, almost close enough to make out the forms of their small bodies.

"Ergon," Mark said nervously, "what are these things that keep coming down the sides of this hole and getting closer to us?"

Ergon glanced down at Mark. **"Those beings are what's left of the civilization on this planet. There are very few of them now. Pity them, and hope your planet doesn't make the same mistakes. Come, we must be on our way."**

The children didn't even have time to blink before the scene below them shrank away and the planet became a dirty tan ball getting smaller in the black sky around it. Again they were streaks of pure energy shooting through the black cosmos. To one side, they could see two blue stars twirling around a huge white one. The sight reminded Katy of a dance she had seen in an old Disney movie, but she couldn't recall why stars would dance like that.

Far ahead, a green dot was growing in size. As they came closer, Katy noticed that the green color was not a healthy green like a leaf or a quiet pond but an ugly, sore-looking green, with a sickly yellow and brown mixed in.

As the group of energy bolts dived toward the surface, the greenish mixture took on the swirling shape of angry clouds before a storm. Ergon took them directly through the horrible clouds that seemed to be miles thick.

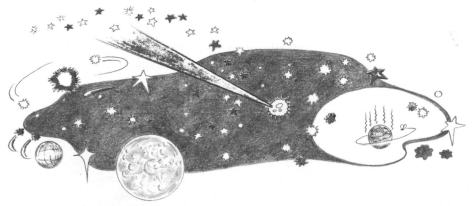

Katy and Mark were returned to human form as the group landed on a small bump of a hill. Around them were smokestacks, buildings, and long twisting pipes that connected everything to everything else. The air, the buildings, and the pipes were the same ugly greenish-brown they'd seen above the planets. The whole place was filthy and disgusting.

Katy said, "This place is a mess!" and took in a breath of hot stinky air. She was suddenly seized by a coughing spasm. She couldn't breathe, her lungs burned, her eyes watered. She gulped for air, but it wouldn't come. Mark, too, was bent over hacking and coughing. Both children were losing control fast, when Ergon made gas masks appear on their faces. Katy and Mark could feel stale, but breathable, air returning to them.

"Sorry. I forgot that you humans need a certain quality of air to live. This planet is Noxious."

Why don't you just let us be energy, like we are when you zoom us around between planets?" Mark asked.

"BECAUSE I WANT YOU TO FEEL THESE PLACES, JUST LIKE THE CREATURES WHO HAVE TO LIVE HERE, YOUNG HUMAN!" Ergon blasted the words out so loud that Mark was shocked into another coughing spasm.

"Just. . .asking. . .that's all." Marked gulped between coughs. Sweat was pouring off of his forehead and down the front of the gas mask.

"It's so hot here, too," Mark mumbled into the gas mask.

Katy had regained her breath and was looking through the eyeholes of the mask at the scene below them. Through the heavy and toxic air she could make out what looked like a shoreline running along the edge of the structure. The water looked thick and oily; all kinds of unnatural things were floating on the greasy surface by the shore.

"What is all that awful stuff, Ergon?" she asked.

"Just that. Stuff," the being replied with contempt. **"You see, these creatures fell in love with all the things energy can produce and never bothered to take care of the chemical and heat waste that is created by the production."**

"Well, where do they live?" Katy responded. None of the buildings or structures she was looking at were fit for anybody's home.

"They can't live on the surface anymore. As you can see, they've ruined it. So they live below the ground in climate-controlled caves, much like your earth animal, the mole."

The group was suddenly transported to a dark, narrow passage, looking out over a large cavern. The air was still hazy and thick, but it was cooler in the cave. Below them, Katy and Mark could see thousands of beings hurrying from one dark hole to another across the floor.

"They look more like rats than moles," Mark said.

"They live like rats, too. Except, unlike earth rats, they can't go outside to the surface of their world unless they're in protective suits."

"Is that a car?" Mark was pointing to a fast-moving object darting in between the scurrying dots on the floor.

"Yes, they have gasoline-powered vehicles," Ergon said with disgust. **"The beings here have everything you have on Earth, even more. The problem is they have never figured out how to have the things they want without creating huge amounts of waste. But they can't hide underground from the pollution that's killing the planet. This world is drowning in its own by-products."**

"What are 'by-products'?" Katy asked.

Back on the surface of the planet again, Ergon continued, **"When raw materials are made into products there will always be some waste energy and matter left over. These wastes are usually unwanted chemicals and heat. Rather than think of them as waste, which means they don't have much use, I like to call them by-products, which means that they could be made useful, or at least harmless, with an intelligent effort."**

Katy squinted through the dense air and saw objects in the water that looked like cardboard, old machines, and trash. On the ground below them she saw glass, cans, and paper. She mentioned each item to Ergon.

"That's all waste to them," Ergon answered, **"but that's the obvious stuff. What do you see in the air itself that's a by-product?"**

Heavy clouds of smoke were climbing into the ugly sky from the smokestacks. By the shoreline, the waters nearest the factories were bubbling

in a poisonous-looking brew. There wasn't a tree, or even a bush, in sight.

"Are those clouds of exhaust and that junk going into the water by-products?" she asked.

"Yes. Those are unused by-products of production, different forms of pollution, poisoning this world. Let's make you two molecules so you can see how the planet reacts to the waste."

Ergon gathered the tiny forms and flew them to the top of one of the giant smokestacks. Mark and Katy were tumbling in the air, rising with the heat of the smokestack and bouncing against one another and other molecules.

"Hey," Mark shouted, his round, tiny body turning and flipping over. "I was happier flying through the galaxies! I get carsick! Let me down." But no one heard or cared about his complaints. Within a few moments, the Mark molecule and the Katy molecule had risen through the thick ugly green cloud that covered the planet. Up at the top of the cloudy mass, the sky was clear and stars could be seen in a blue-black sky.

Ergon's voice seemed to come from a distant sun.

"You are now part of the carbon dioxide gas, or CO_2, layer surrounding Noxious, the planet below. The problem is you are so thick, so dense, that you have blocked the normal escape of heat from the planet's surface. Your gases act like a mirror to the heat trying to get out. As a result, the planet is getting hotter and hotter."

"I know what that is called," Katy declared. "That's the greenhouse effect. Earth scientists argue about whether it's really getting worse or not."

"The greenhouse effect is not bad. As a matter of fact, it is necessary for life on a carbon-based planet. What's bad is when chemical pollutants in the air increase the greenhouse effect and raise air temperatures to unnaturally high levels. That's what has happened here and what may be happening on your Earth."

"You know, it's nice and cool up here at the top," Mark said, bouncing away in the distance. "I wouldn't mind hanging around here for a while." As he was enjoying the ride, he suddenly started to feel heavier. The air around him became thicker, and he began falling into the cloud layer.

"What's happening?" Mark shouted. "I'm sinking fast. Help!"

Through the confusion of millions of other molecules falling downward,

Mark could hear Katy yelling, "Me . . . too!"

"Beware of another effect of pollution in the air!" Ergon's voice cut through the rush and fury.

Suddenly, Mark was attacked by ice crystals. The jagged riders seemed to grab onto him and not let go. Then he was a small chunk of ice, still falling and gaining speed. As he broke through the bottom layer of thick cloud cover, he turned into water. He was rain, and he was dropping toward a scrubby-looking hillside of tree skeletons. Within the water droplet that was Mark there was something else. He could feel its presence as an alien thing, something that didn't belong with the water part of him. This must be what Ergon said to beware of, he thought.

With a hard PLOP Mark hit the needles of a nearly dead pine tree. Mark, and the water part of him, slid down along the pine needles and fell to the ground, but the strange part stayed behind and clung to the tree. Mark was glad to be rid of it.

"The alien thing you felt was the acid in the rain. The one that attached itself to you is sulfuric acid, H_2SO_4. It is another by-product of careless energy use and production. Acid rain kills the very trees and plants that could help the planet recover."

Katy and Mark had both returned to human form and were looking around the sad, dying forest.

Ergon paused, and his pulsing body seemed to lose a bit of its glow. **"So, here on Noxious you've seen the worst results of making products and energy without regard to controlling the by-products, or waste. A planet can strangle in its own material wealth."**

The children were just beginning to cough and wheeze again as Ergon said, **"We have one more stop to make before we head back to Earth."** In an instant, the group was flying away from the terrible planet with the deadly blanket.

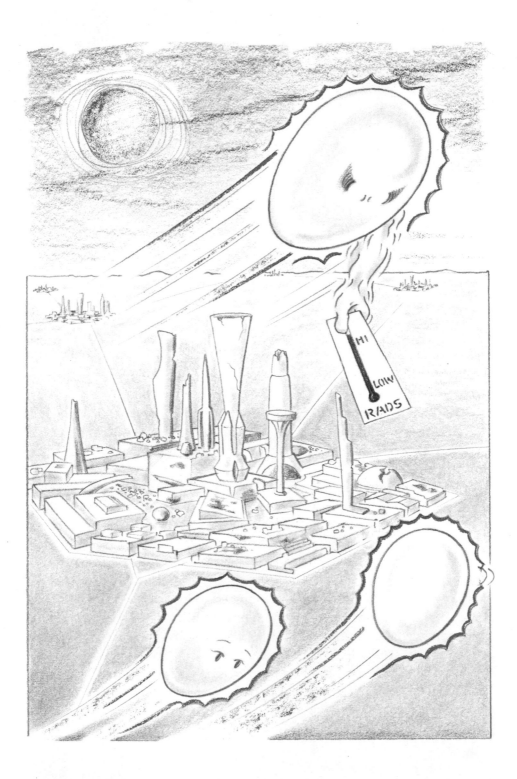

3

DEAD END

Through the black of the cosmos they sped. They passed galaxies shaped like Christmas ornaments and some that looked like shiny coins. Katy tried to think of the beauty and wonder around her, but her mind kept going back to the survivors on Barren and Noxious.

Up ahead, a gray ball was growing larger. As it grew, Katy could see that it looked like stone, with no blue or green colors that might show signs of life in oceans or forests. Ergon was guiding them down to the planet, but they weren't approaching the surface as they had when they visited the other planets. The group leveled out and began flying over the ground just low enough to see groups of buildings and roads that looked like they were once fantastic cities. All the buildings were clumped together to form a pentagon, or five-sided figure, with long straight roads connecting one city to another. Nowhere did Katy or Mark see any movement of vehicles, a green field, or even a lake, just a continuous grayish-silver land with a few mountains breaking up the surface.

"Does everyone live underground here, too?" Mark asked Ergon.

"No one lives here at all," Ergon replied. **"Anymore."**

"What is . . . I mean, what was this place?" Mark continued.

"Sometime long ago it got the name Dead End."

"That's a weird name," Mark said. "Are we going to go down to the ground to look at the place?"

"I don't think so." And for a moment, Ergon paused. **"The amount of rads down there might penetrate any protective shield I could put around you."**

"What's a rad?" Mark said with excitement. "Some kind of horrible monster?"

"You might say that," Ergon replied dryly. **"Actually, a rad is a measurement of radiation and how it affects health. Too many of them and you would get sick and die. Dead End has more than enough rads to kill any life-form."**

Katy knew that something awful had happened on this planet and thought for a while before she finally asked Ergon, "Can you tell us about it?"

"It's a sad story, but one you should know." Ergon began telling the tale. Early in the planet's history, the beings discovered the great energy in nuclear power. At first, this power was used for weapons: one nation on the planet would use the weapons to bomb another. The planet was lucky to survive the war period. Finally, peace was declared and all the atomic weapons were taken apart. The atomic material from the weapons was kept in different places around the world.

Next, ways were found to use nuclear energy for making electricity, powering vehicles, and running factories. At first, it seemed to be the perfect solution. Nuclear energy had little visible pollution, and a few small and rare mines could provide all the raw materials ever needed. The by-products of the process were deadly but didn't take up much space. The people had all the power they wanted and thought the world was fairly clean.

"It sounds so good. What could have gone wrong?" Mark wondered out loud.

"Several important things were going wrong. First, no matter how much thought and care went into the design and building of the power plants, they began to break down every so often. When the accidents were major, damage to the people, animals, and microorganisms for miles around was huge."

Katy and Mark both remembered hearing and reading about similar disasters on Earth. One area of the USSR was still closed to everyone and all human

activities within a large zone had been stopped. The land was poisoned. The people who lived there when the accident occurred had terrible diseases.

Ergon went on with the story. **"Another problem was the leaders and scientists of the world could not figure out a way to get rid of the small but deadly amount of radioactive wastes left over after the nuclear fuel was used.**

"They tried putting it in cans, concrete, ceramics, all kinds of containers, but the deadly waste would get out. Next, the scientists decided since the stuff was so dangerous and escaped from anything they put it in, the waste should be buried underground. The problem was where to bury it."

"Why not just take it out to a desert somewhere and bury it there?" Katy asked. "It couldn't hurt anyone if no one's around."

Ergon explained that by this time there were enough people in the world so that even in the loneliest deserts, someone was close by. Besides, people were worried about other things. Some were worried about getting the waste to the storage places in the deserts. In a wreck, the radioactive material could always spill out or leak. Still others were worried that the deadly waste would eventually get out of the underground storage areas and seep down into the water tables that ran below the ground, ruining the water for cities far away.

"Those are a lot of things for people to worry about," Mark said quietly.

"Some of the beings kept their sense of humor. They suggested

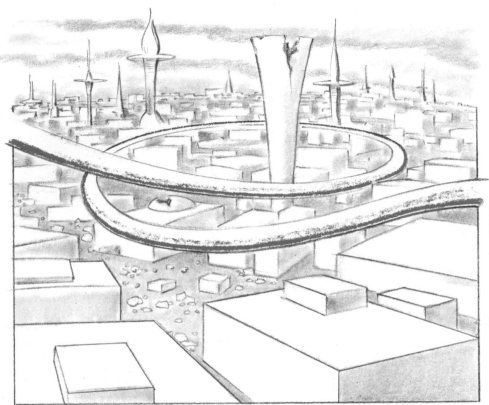

that if the waste was so safe, the leaders should bury it in their own backyards."

"Oh, yes," Katy chuckled, "I remember what you said earlier about treating all the planet like your own backyard."

Ergon had taken the group a little lower, and now they were circling over the largest pentagon-city they had seen on the planet. It stretched from one side of the horizon to the other. Scattered across the landscape were the ghost towers of skyscrapers and squat cubelike buildings that covered entire parts of the city.

"The most serious problems concerning Dead End's energy future didn't occur to the beings until it was too late," Ergon said. With so much importance being placed on making nuclear energy safe, the planet didn't develop other options. Many of the best energy scientists were trying to find safe ways to generate, store, and transport the atomic power. Others were looking into new ways to create nuclear reactions. The world govern-

ment put too much of its brain power and money into finding these solutions, and more problems and breakdowns were happening all the time.

"**After all,**" Ergon concluded, **"even the richest world has many other problems besides power, and there's only so much money to go around."**

"My grandmother would say that they put all their eggs in one basket," Katy remarked softly.

"Yes, and they dropped the basket."

All this time Mark had been quiet, which was unusual.

"Ergon, does one of these energy futures have to be ours?"

"DEFINITELY NOT! Now that you know what the future could be, you must gain the knowledge to change it to what you want!"

The gray, deserted planet of Dead End fell away as Ergon pointed the group up into space and the glittering light of a million stars.

4

A HEATED DISCUSSION

ark thought they might be getting close to home when he spotted Jupiter. It looked just like the old Voyager poster on his bedroom wall at home, only the colors were brighter. The Red Spot was so large and angry-looking he could almost feel the force of it as the group passed. In no time, Mars could be seen clearly in the distance, orange and dusty with a white polar ice cap as a hat.

"**Your solar system, and your star Sol,**" Ergon said.

The sun, straight ahead, was growing larger. It was about the size of a dime when Mark spotted his Earth (that's the way he thought of it now). Soon Earth began to show the soft blue colors of the water and the white of clouds.

"**This is the star responsible for all forms of life on your planet,**" Ergon noted, as the sun began to show yellow and orange in its body.

"Hey, Ergon," Mark called. "It's the same color as you when you're not mad!" Right away, he was sorry he said it because the energy being beside him began to glow redder.

Katy voiced the feeling many astronauts had when they returned from a voyage into space. "Earth looks so fragile and beautiful from up here."

"**It's both of those things. The pictures of your planet taken from space made people aware that the world is basically a closed system.**"

"What does that mean?" Mark asked.

Ergon told him that in a closed system, everything was linked to everything else. An action taken on one part of the system would change another part, even if the two didn't seem to be connected.

"I don't follow what you're saying," Katy said.

"**Maybe the best way to show you is to start at your house.**"

The children were happy and excited to be going home, although they had no idea if this was the end of their trip with the strange energy being or just the beginning of a new one. Soon they were diving down through fluffy white clouds, and Mark could see a seacoast approaching quickly across the blue choppy ocean. Towns and cities blurred underneath them as they raced across the landscape. Mark thought he spotted a familiar water tower and park. The next thing he knew, the group dove through a roof and Katy, in human form, was standing beside him in their living room.

"Whew. Great trip," she said. Katy felt relieved seeing the familiar items of her house around her. She also noticed that there was only one small light on in the room. Evidently, Ergon had turned off all the lights and appliances when they made their hasty departure.

Out of habit, Mark was reaching to turn on the television when Ergon's voice stopped him just as his hand was touching the power switch.

"Before you turn that on, would you like to know what happens to energy as it does different jobs in your house?"

"Sure," Mark replied. "If I can get back in time for my favorite show."

Ergon began to glow redder again, and a low throbbing noise seemed to come from deep inside his form. Right away, Mark added, "Really, it doesn't make any difference when I get back. I was just kidding about the TV show." The creature seemed to like that answer better, and his color went back to a pleasing soft orange.

"Let's start here," Ergon said. **"Katy, go over to that small light and carefully put your hand over the top of the shade. Tell us what you see and feel."**

Katy followed the instructions and placed her hand gently over the lamp.

"Of course, I see light from the bulb (although not enough to suit me in this dark house, she thought to herself). And my hand feels warm from the heat of the lamp. That feels good, because this house is a little cold." It seemed to her that Ergon must have also turned down the thermostat when they left for their voyage to the three planets.

"So, you are seeing and feeling two forms of the same energy, light and heat. Which is the main product, and which is the by-product of the lamp?"

"Well, a lamp is supposed to make light, so I guess the light is the main product and the heat is the by-product. But the heat is welcome, too," Katy said.

"Would the heat from the light be welcome in August in this house?"

Mark answered, "No! It sure wouldn't. It gets hot here in the summer, and we have to use the air-conditioning to keep the house cool." Katy agreed with her brother.

The lamp is doing two jobs, Ergon explained, lighting and some heating. Then the being told the children, **"All machines have heat as a by-product of producing the energy humans want. Sometimes, like with the light right now, the extra heat is needed and used. Most of the time it is wasted and goes into the atmosphere."**

"But what about refrigerators and air-conditioners?" Mark asked. "They keep us cool. They make cold air, not heat." Mark felt that he had caught Ergon in a mistake and was going to rub it in.

The next instant, Katy and Mark found themselves in the kitchen sitting on top of their refrigerator. Katy's head was bumping against the ceiling, and Mark's feet were dangling over the edge of the door.

"Put your hands carefully behind the back of the machine. You don't have to touch anything. Just tell me what you feel in the air behind it."

Both children did as Ergon said. It was Mark who spoke first. "It's warm back there!" The creature explained that cooling machines all have heat as a by-product and that air-conditioners, particularly in big cities, make the air in the city warmer even as they are cooling the buildings. Many air-conditioners also use chlorofluorocarbons, or CFCs, which can shrink the ozone layer protecting and surrounding the earth. **"Here's an example where one type of energy used, cooling, creates a larger problem in the world's environment,"** Ergon said.

As Katy and Mark jumped down from the refrigerator, Ergon challenged them to find as many machines as they could in the kitchen which made waste heat.

Mark ran immediately to the gas stove and oven. "Here," he said. "This is one." As he looked at the stove, he noticed the hood and light above it. "There's a fan in here, and it takes the cooking smells and, I guess, the leftover heat from the stove, outdoors." He opened the oven door and noticed that his pizza was cold and dry. He was tempted to sneak a bite when Ergon wasn't looking but changed his mind and shut the oven door.

Meanwhile, Katy was standing between the toaster and the microwave oven. "I understand that the toaster makes heat from electricity, but how does the microwave get food hot?" she asked.

"Electronic waves excite the molecules in the food and all the

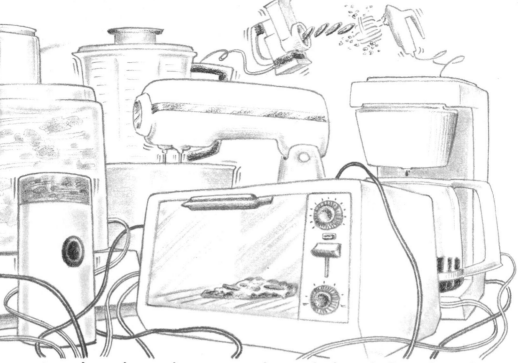

atoms bumping against one another make it cook. But where does the waste heat go?"

Katy opened the door and felt the barely warm bagel she had been heating before Ergon's arrival. "Well, I guess the heat leaves the food and goes into the air in the kitchen," she said uncertainly.

"That's correct." Ergon paused for a moment, and both children looked at him intently. **"Heat always goes to a colder area. This is an important law of nature you should understand if you want to discover how your use of energy affects the Earth."**

Mark had opened the door to a small closet in the kitchen where the water heater for the house was located. As he did, he could feel the heat escaping from the tiny room. "Boy, it's hot in here," he said.

"That's one of the biggest heat losers in your house," Ergon replied. **"Most of the waste heat goes out of the walls or right up that pipe you can see going through the ceiling."**

The children were looking around at smaller machines in their kitchen. Katy spotted the blender and the food processor on the counter.

"These don't have waste heat, do they, Ergon?"

In answer, both machines went on with a loud whirring, and Katy jumped back in surprise. Ergon told her to wait until the appliances had run a bit and then to feel around the base of the machines. She did and was amazed to notice

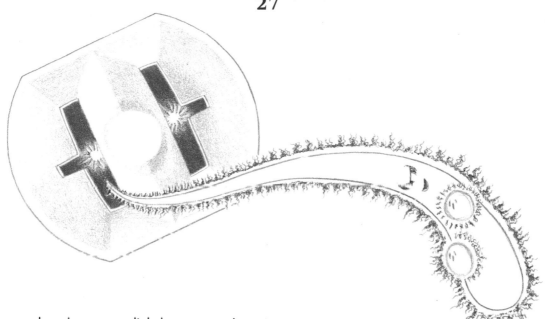

that they were slightly warm and getting warmer.

"That's waste heat caused by friction, the parts of the machine rubbing together inside of it. You could say that the hotter a machine gets from friction, the less efficient it is and the more energy it is wasting."

"Efficient? What does efficient mean?" Mark asked.

"Efficiency is a way to measure how much energy goes into a machine or a system and how much work comes out of it. The more efficient something is, the less energy is wasted in heat or in other by-products that become waste. The electrical power supply is a giant system supplying millions of machines throughout your country. The work that comes out of it, you see here in your home. To find out what goes into it, we have to cruise the currents."

"Cruise the currents?" Katy and Mark said together.

But Ergon was changing shape now, shrinking and growing smaller. So were Katy and Mark, as the kitchen counters around them began to look like huge cliffs. The next thing Katy knew the wall socket by the toaster was rushing toward her. She was so small that the electrical inlets in the socket looked like long narrow black caves, and she was zooming toward the one on the left. Katy could hear Mark yelling in the background.

"You could get a bad shock doing this."

5

CRUISING THE CURRENTS

Into the electrical outlet they flew. For a moment, they charged around inside Katy and Mark's house, then ended up at a large gray box.

"Junction box," Ergon said, as they quickly went through the device and out of the house. **"That's where the electricity enters your house."** Already they were out at the street and into the wires hanging from power poles. Katy looked toward Mark and found that he was like her, a tiny but brilliant spark. They were following along behind Ergon, who was in the same form. A block up the street, the group of sparks traveled around inside of a larger gray box and then they were back in the wires. Katy noticed that she felt warmth in the large box and stronger as she came out.

"Ergon, where are we going this time?" Mark asked.

"We're following the path that electricity takes to get to your house. We're in the system moving toward the source of electricity. You can imagine yourselves as free electrons. That's how electricity moves."

"But we're going so fast," Mark said. "Look, there goes my school and now we're almost out of town."

"Actually, I've slowed things down a lot," Ergon replied. **"If we were going as fast as electricity travels, we'd be going the speed of light, and you wouldn't be able to see anything."**

"Ergon, why did I feel warmer in the gray box on the pole outside my house?" Katy asked.

"That was a pole transformer. It takes the electricity from the system and directs it into your house. Some energy is lost in the process. That was the heat you felt. You"ll feel stronger as we go up the lines, because the electricity is more powerful as we get closer to where it is generated, or made."

Now the wires were entering a complex of strange-looking gray buildings and machines near the edge of town. Mark had noticed the site many times. It looked like the bumps on the outside of a movie spaceship. The entire area was surrounded with a high chain link fence and a bunch of signs saying, "HIGH VOLTAGE" and "DANGER—KEEP OUT." Again, the children felt warm as they zipped through the group of objects.

"This is the local substation for your community. These machines transform high voltage electricity, too powerful for your home, into lower voltage and send it along to homes and businesses."

When they came out the other side, the group shot up the wires high into the air and onto metal towers. The increase in their feeling of power was huge. Now, all around them was a crackling and hissing of power as they passed high above fields and farms.

"I know what these are," Mark said. "These are high voltage power lines. Some people on a farm outside of town tried to stop the electric company from putting these through their property."

"The electrical power in these lines is tremendous," Ergon answered. **"Energy loss through the lines happens in the form of heat and electricity lost into the atmosphere."**

Every mile the group traveled, the children felt stronger and stronger. Ergon explained that they were now in a "grid" that could take them just about anywhere in the United States. The grid lets power companies share the electricity throughout the country, so if one area runs short of power, the power company can borrow electricity from another company nearby. This sharing of power is called a "pool."

Now the group of electrons were approaching a cluster of buildings larger than the previous one and off to the side of the power lines. It had a big smokestack in the center and another unusual-looking building connected to the lines.

"The building in the center is a small oil-fired generating plant. Oil, like coal, is a fossil fuel created from the bodies of plants and animals millions of years old. Because oil costs more than coal, this plant is only used at special times when electrical needs are great."

As they got to a split in the power line, Mark shot off in the direction of the small power plant.

"COME BACK HERE NOW!" Ergon ordered. Mark pretended not to hear and was headed down the lines toward the closest building. Making a loud crackle and a flash of light, Ergon went after him and caught him in an instant.

Mark was looking kind of gloomy as Ergon brought him back up the line toward Katy. **"DO YOU KNOW WHAT COULD HAVE HAPPENED TO YOU BACK THERE? YOU COULD HAVE GONE THROUGH THE GENERATORS, PAST THE TURBINES, INTO THE BOILER, AND BEEN TRANSFORMED INTO A HEATED WASTE CAR-BON MOLECULE! THAT'S WHAT!"**

As an electron, Mark was useful to humanity in electrical form. As a heated carbon waste molecule, he would go into the atmosphere, team up with an oxygen molecule, and form carbon dioxide. The earth had too much of that already and didn't need Mark to add to the supply.

"Remember the planet Noxious?" Ergon asked. **"That's exactly what happened there."** Both Katy and Mark were saddened as they thought about the planet choking in its own waste.

If an electron could crawl under a rug, Mark would have. "I'm sorry," he said. "I won't do that again."

"Just to make sure it doesn't happen again, I'm going to make Katy a positive charge (+), and Mark, you'll be negative (–). Now try going away on your own."

Katy and Mark tried to pull away from each other, but a strong force kept bringing them back together again. Ergon told them that electricity could take several forms, including magnetism, lightning, and chemicals, but it always required positive and negative electrons to work.

Ergon explained to the children that he was going to show them how the electrical generating process happened but that they had to stay together and let him guide them. He reminded the children that they were moving back-ward, from the end use in their house to the original source, fossil fuel. The process involved changing the energy in the form of carbon-based fossil fuels, like coal, into electrical energy. It involved several different steps. During each step, energy was changed from one form into another. Every time the form of energy was changed, something, either heat or chemicals, was added to the environment.

The group accelerated again up the electrical lines. In the distance they could see a huge group of buildings, power lines, smokestacks, and great piles of materials. The group passed through a jungle of wires and towers and entered the largest building. Throughout the huge space were giant machines, row upon row of them.

"These are the generators that turn one form of energy, in this case, mechanical energy, into another, electricity."

The children found out that power plant generators change mechanical energy into electricity through the use of magnets and magnetic fields created by spinning parts within the machines.

"Magnets, just like the one I got for Christmas?" Mark asked.

"Just the same, only much bigger," Ergon replied.

The important thing, Ergon said, is that generators need an outside force, mechanical energy, spinning the parts of the generator so they can make electricity.

Ergon took the group down the wire and into the generators. They were still electrons, but instead of being in a wire, they were inside a big coil of copper within the large machine.

"This is the point where electrical energy is made." Katy and Mark could hear Ergon's voice above the crackle of billions of electrons forming.

"This part of the process is very efficient. Almost all of the mechanical energy becomes electricity." But now, Ergon was moving them toward the center. **"You're about to change form again!"** When they got to a large spinning shaft in the middle of the machine, their structure began to change. From brightly sparking electrons they were changing to objects that looked like tops, little wooden tops spinning at a tremendous rate.

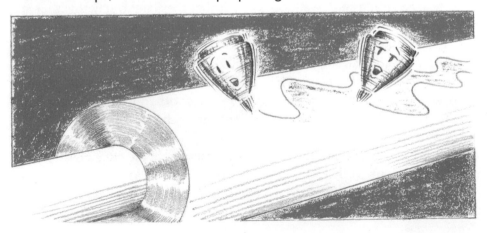

"Now you are mechanical energy," Ergon said.

"Whoa!" Mark shouted. "I liked being electricity better. I got a charge out of that. This could make you real dizzy in a hurry."

"In a power plant, you won't be mechanical energy very long. You are needed in this form just long enough to turn these generators." Ergon took the group off of the spinning shaft so they could see the layout of the entire area and the machines that turned the shafts. Mark didn't like the looks of it one bit.

"Wow," Mark said. "Are you sure we ought to go into this one? It looks like a giant jet airplane motor." Ergon said that the motors are similar and work on the same principle.

"These are turbines. They produce mechanical energy that turns the shaft you were just on."

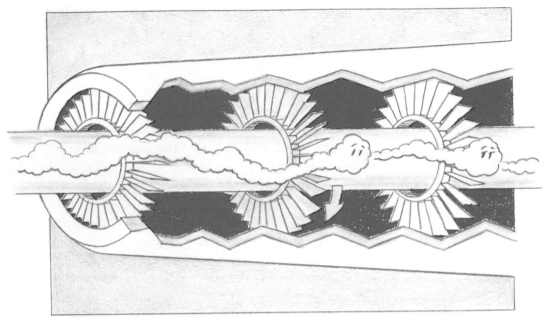

Large pipes could be seen entering the far end of the turbines. The spinning shafts were coming out of the closest part and going into the generators.

Katy said, "By the temperature in this area, I can tell that a lot of energy is going into the atmosphere."

"That's right," Ergon said. **"Many of these turbines are old. To make power, they have to work harder than new and better machines, so they produce extra friction and heat when they could make more electricity."**

Katy and Mark learned from Ergon that the United States could save as much as *25 percent, or one-fourth, of its power* if old motors of all kinds were replaced or fixed with new and better parts. These motors could be small ones in homes, like electric knives and blenders, as well as big ones, like the motors they were seeing in the power plant. Motors of all kinds could run easier, or more efficiently, and there would be great energy savings for the world.

"Today, power companies try to capture waste heat from electrical generation for useful purposes. But a lot still escapes into the environment," Ergon explained. The children were beginning to enjoy their new roles as tops, and Katy was spinning along the edge of one of the turbines.

"Mark, look at me," she said, "I'm the world's greatest ballerina."

"ENOUGH," Ergon called. **"Change of energy form again!"**

Katy and Mark were being transformed from spinning tops to steam molecules, the gas form of boiled water. The group charged into the whirling blades of the turbine. The blades were attached to the shaft, and the spinning is what gave the shaft its mechanical power. Ergon took them through the wildly spinning machinery that was being driven by the power of the steam.

"Gosh," Mark said to Ergon, "you don't have to get so heated up over it." And hot it was. Over 1,200 degrees of heat were needed to get the steam to the right pressure to spin the giant turbines.

Mark and Katy were rolling and bouncing through the pipe against the flow of billions of droplets of steam going the other way.

"There are several ways to turn turbines for electricity. Steam is the most common," Ergon said.

Ergon explained that about 75 percent, or three-fourths, of the world's electricity was made by steam-powered turbines. Most of the heat to make the steam was created by burning the fossil fuel, coal. The waste from the burning put CO_2, and SO_2, sulfur dioxide, into the air.

"But there are other ways to power turbines," he said, **"and I want to show you a couple. To do that, we have to leave this coal-fired plant and travel elsewhere."**

6

POWER TRIPS

The scene below became blurry, and now the group was looking down at a beautiful lake with a large dam at one end holding back the water. Ergon told them that this was called hydroelectric power, which provides about 10 percent of the country's electrical power.

"How does that dam heat the water for the turbines?" Mark wanted to know.

"It doesn't work that way," Ergon explained. **"The dam plugs up a river to form a lake. The contained water, which wants to fall to a lower level, is released through a pipe or tunnel in the base of the dam. The kinetic, or moving, energy in the water turns the blades of the turbine to make electricity. The power generating plant has to be below the level of the water to work."**

"So in this process, there's no waste heat," Katy added. "It must be very clean."

"Actually, there's some friction heat loss, but not much. It is a very clean way to make electricity. The power of the falling water makes electrical energy from mechanical energy without going through the heat step," Ergon said.

"Oh, it's beautiful," Katy said. Below them there was no ugly brown smoke coming from the power plant. The children saw sailboats and the wakes of other boats pulling water-skiers. The scene was pretty, especially compared to the coal-fired plant they had been cruising.

"Why don't we make all our power this way?"

Ergon told Katy to think for a bit about any problems she could see with making new dams and more hydroelectric plants.

"Well, they cover up land when the river backs up," Katy said, after some thought. "Is that right?"

Ergon said that was correct. Both farmland and scenic land were valuable, and people often didn't want it to be underwater. Ergon told a short story about a large dam in the western part of the country that covered up one of the most beautiful canyons in the world.

Mark added, "I guess there are parts of the country where there isn't enough water to do this." That was true, Ergon replied. Other places are too flat, with no natural cliffs or barriers to hold the water along the sides of the new lake.

"What about the effects on the river below the dam after it is completed?" Katy said.

"That's important," Ergon said. **"And something more humans are becoming concerned about. Once a river is dammed, it is changed. Fish, wildlife, and plant life below the dam will never be the same. New kinds of life will come in, and old kinds will die."**

Ergon wanted Katy and Mark to look at another form of power generation that didn't rely on heat. The scene below them changed. Now they were looking down on a flat valley high in the mountains. They could see hundreds, maybe thousands, of small towers with propellers on each one. The propellers were turning round and round and, from above, looked like the old-fashioned pinwheels you can buy at a circus or fair.

"What's moving all those propellers?" Mark wanted to know. "It must take a lot of electricity to make them go like that."

Ergon chuckled. **"Mark, the propellers are making electricity from the wind, not the other way around."**

Mark was confused. "How can that happen? There's nothing to run on but air."

Wind, Ergon explained, carries huge amounts of kinetic energy all over the surface of the earth. That power is often felt in tornadoes, hurricanes, and thunderstorms in a negative way to humans. Just the normal everyday breezes and the powerful jet stream flowing several miles above the earth contain more energy than the world can use in a day.

"Well, that's the way to do it then," Mark said happily. "Let's use the wind to make all our electricity. I can tell there's no pollution or waste heat here. Sometimes the wind at our house blows fifty miles an hour. We could just put up a giant tower and . . ."

"Hold on a second," Ergon said. **"Wind power is a great way to get electricity, but it does have some drawbacks."**

First, Ergon explained, the wind needs to blow with a constant, steady force.

Unfortunately, many parts of the Earth don't have a constant wind through-out the year. They have storms and windy seasons when the wind blows like crazy but not the day-to-day dependable breezes a wind turbine needs. When such conditions are found, often on seacoasts and mountain passes, the people who live in the area have to agree that all the machines aren't causing "visible pollution." In other words, the local people have to like the looks of the wind machines or they would never be happy with them.

"Well, they look a lot prettier than a regular power plant to me!" Mark said.

"Unfortunately, hydro and wind power don't make very much of the electricity your planet uses. It's time we go back."

Before Katy or Mark could enjoy more of the view, they found that Ergon had returned them to the boiler of the coal-fired plant. The heat was intense, and they were tumbling around with Ergon in the inferno.

"Stay close to me!" Ergon's voice was clear above the thundering sound inside the boiler. All around them, coal particles were catching on fire. The kids could see particles ignite, but a large number didn't burn. These were carried upward toward the large smokestack at the top of the container.

"What are those things that aren't burning, Ergon?" Katy yelled.

"Those are different kinds of matter that don't ignite in here—sulfur, carbon, and other elements that escape the fire to rise into the outside air."

"Let's get out of here," Mark screamed. "This is too dangerous, even for an energy creature." The next thing he knew, the three of them were on a moving belt headed down a long ramp. It was strange because all the coal was passing them heading up the belt to be dumped into the boiler. Ergon told them this was the conveyor belt that took the coal from trains and trucks and deliv-ered it to the boiler.

It was quieter now, and Ergon reviewed what had happened. **"Here at this plant, you see the way 75 percent of the world's electrical power is made. The coal partially burns, causing high heat. The heat boils water, turning it to steam, a gas. The steam turns the turbines, making mechanical power in the shaft turn the generator, which makes the electricity."**

"So, our electricity all comes from boiling water?" Katy asked. That was true, most of it did, Ergon answered, except for hydro and a small amount of wind-generated electricity. Above them, the children could see two giant smokestacks and the unburned wastes going into the air.

"Why can't something be done about all that junk going into the environment?" Mark wanted to know.

Ergon answered that there are ways of taking the waste particles out of the exhaust. One method is to install a device on the stack, called a "scrubber," which cleans many of the pollutants out of the smoke. Power companies were adding more scrubbers to their plants, but they cost a lot of money. Another way, Ergon said, is to use cleaner coal in the first place. Those ways add money to the production of electricity and would show up as higher electrical bills for homes and businesses.

"I think people will pay a higher price for cleaner air," Katy said.

"They might not have to pay more if they would start being energy smart." The children didn't understand, and said so, but Ergon chose not to answer that question at this time. Instead, the energy creature informed Katy and Mark that they had one more stop in their tour of power plants.

This time, they flew over the land a short distance and stopped on a hillside overlooking a large bay and power plant. Katy and Mark could tell this was not a coal-fired plant because of the jumbo-sized fat tower. From the top of the tower, the group could see a constant stream of white smoke that went for a little way into the air and then disappeared.

"See, more sulfur and carbon waste," Mark said with disgust.

"Wrong," Ergon said. **"Guess again."**

"I think it's steam," Katy said. "Because it only goes a little way into the air and then it's gone. It's a nuclear power plant. Am I right, Ergon?"

"Yes, you're right."

Meanwhile, Mark spoke up. "If it's just steam, then it isn't carrying any of the pollutants of the coal-fired plants, is it? So it must be a better way to make power."

"How do you suppose the electricity is made here?" Ergon asked Mark.

"Well," Mark started, "I think the atoms in there bounce around and create an electrical charge and go into the lines as electricity."

"And where does that happen?"

"Probably in that fat building over there, the one that's shaped like a golf ball," Mark said.

"Sorry, but you're wrong, Mark," Ergon said. He went on to explain how a nuclear plant isn't that different from a coal-fired plant. The whole effort is to get the temperature of water high enough to make super-hot steam to drive the turbines. To do this, an atomic reaction had to be controlled night and day so that it wouldn't overheat and "go critical," or get out of control. The process demands constant watching and care. Waste heat from the water used to cool the equipment meant that a nuclear power plant could cause serious changes to the surrounding environment if hot water was dumped into rivers and bays.

Ergon told the children how for over forty years, governments and power companies had spent billions of dollars on this form of energy and that it still had many problems. At one time, he said, the whole world thought this would be the energy for the future. But after all the time, money, and brainpower that went into it, atomic power only provided the planet with a little over 10 percent of its electrical needs, about the same as hydroelectricity. Now a great amount of money and scientific effort was going toward finding safe ways, if there were any, to get rid of leftover radioactive wastes and decommission, or shut down, the old plants.

"So all this whole business does is *boil water?*" Katy asked.

"Basically, yes," Ergon said. **"One of your humans has said that using nuclear energy to boil water is a little like using a chain saw to cut butter."**

"It seems like a lot of effort for a little result," Katy said.

"Looks like a dead end to me," Mark added.

Ergon seemed to smile as he whisked the children into the air and away from the power plant. **"There are important uses for atomic energy — medicine, space exploration, to name a couple. Boiling water to make electricity is probably not one of them."**

7

GOING WITH THE FLOW

Katy, Mark, and Ergon were back on the conveyor belt at the coal-fired electrical plant. They watched as machines below unloaded coal from open train cars, placed it on the moving belt to be lifted, then dropped it at the boiler. Ergon spoke, **"You've seen a great amount. Tell me what you've learned. What about the distances electricity has to travel?"**

Mark was uneasy, but he began, "It's a long way from the electrical outlet in our house to the power plant where the electricity is made." Katy and Mark both knew now that the longer the distance, the more electrical power was lost along the way to transformers and line losses. Mark told Ergon that shorter distances meant more usable electricity at the end.

"How about the steps in producing the power?"

"Well, energy has to change forms several times to make electricity," Mark continued. "Every time the form of the energy is changed from chemical to mechanical to electrical, some of it is lost to heat or waste."

"I've learned that most of our electrical power comes from boiling water to make steam," Katy added. "And what is used to boil the water often creates the pollution and safety problems."

"And that brings us back to here and the coal." Ergon paused and then continued. **"As your planet is so dependent on the fossil fuels, coal and oil, I think it's time that you visited one source of your planet's wealth."**

"Are we going to some bank?" Mark was saying, but his last words were lost in the rush of wind as Ergon took the group high above the land. Night was falling across the continent: to the east it was almost black, while in the far west there was still sunlight. Off to the right side, the children could make out the shape of the Great Lakes as they headed past them in a northerly direction.

"Where are we going now?" Katy wanted to know.

"To one of your country's greatest areas of natural resources, Alaska," Ergon answered. But already the group was coming down. They passed over great forests that seemed to go on forever, with high white moun-

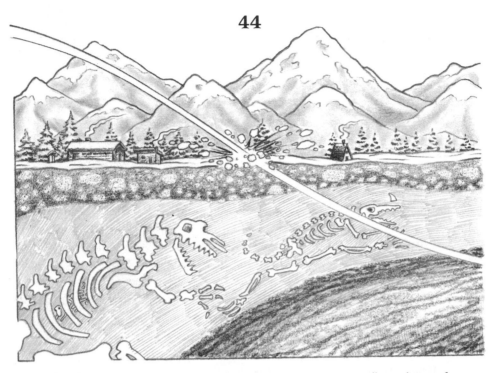

tains breaking through the trees. They flew lower over endless plains of snow. Herds of elk and lonely bear were passing below them. Ergon took the group down through low clouds and toward a settlement of scattered buildings and large groups of machines.

"Hold on . . . Ergon . . . you're going to crash us into that . . ." But before Mark could finish, the group had gone through the surface of the Earth and was headed deeper into the blackness.

"Mark, you must remember that you're not in human form when we travel like this," Ergon said. **"It will save you a lot of worry."**

But Katy was feeling some of the same alarm. The total blackness and closed-in feeling she was getting was not comfortable. The temperature was getting warmer as they went farther into the hole. Down they dove, deeper and deeper into the earth. Just when Katy was about to express her alarm, Ergon began to slow their fall and brought the group to a slow stop.

All around them was a strange sensation of great age and time. Mark noticed it, too, and said to Ergon, "How come it feels so old down here?"

"How old does it feel to you?" Ergon answered Mark's question with another.

"Ancient," Mark replied. "Maybe one hundred or two hundred years old."

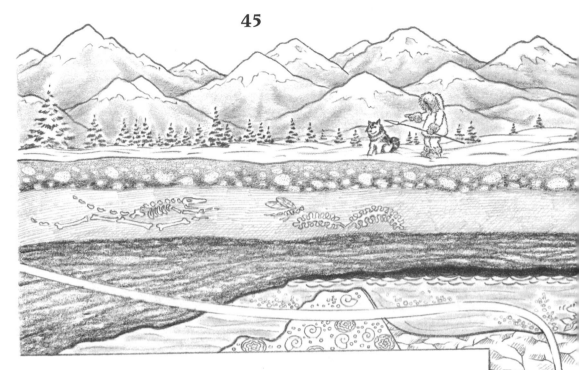

"TWO HUNDRED YEARS? WHY DON'T YOU ADD SIX ZEROS TO THAT? YOU'D BE CLOSER TO THE ANSWER."

Two hundred million years—the children couldn't really imagine such a long time ago, but they knew it was long before humans walked the Earth. Ergon calmed down and explained that scientists believe that millions of years ago, oceans covered much more of the Earth than they do today. During that time, it was probably warmer than it is now on Earth, and there was more plant and animal life in the seas and oceans. Over time, these trillions of little creatures died and were covered by sand and silt from the oceans. As the Earth's crusts moved and changed, the great pressure and weight of the land above them compressed and squished them into the oil, natural gas, and coal that Earth depends on today. The reason they are called "fossil fuels" is that they are actually made from living things.

Ergon paused and then said to the children, **"I'm going to let you feel some of that pressure. It won't hurt you, so don't panic."**

Katy and Mark began to feel a tremendous pushing from all sides. The weight of the rock above was matched by an equal pushing up from below as the great forces in the Earth fought with one another. Mark was being squeezed right into a rock, through the tiny passages within the stone. He found himself, along with millions of other oil droplets, buried inside of the rock itself and moving very slowly through it.

"That's one way oil is found, Mark," Ergon said without much concern. **"As oil, you'd continue to move that way for millions of years until you reached a rock that couldn't be entered. Then you'd either form a pool of oil trapped by the denser rock or you might stay in the softer rock forever."**

"I don't want to do either!" Mark shouted, but Ergon was now with Katy. She had gone a different direction and had been pushed up from below by water in the earth. Katy had also taken a different form. She was lighter-weight natural gas, found above many oil deposits.

Katy, along with her gas molecule neighbors, was covering the top of the oil pool and forced against a solid rock cap above. She couldn't go anywhere.

"Hooboy," she said, "this is worse than the cafeteria at lunchtime. Everything's squeezing in on me."

Ergon let both kids remain in their petrochemical state for some time before he brought them back to his side in the oil pool.

"There are several things here I want you to remember," he said very seriously. **"First, oil, natural gas, and coal took millions of years to make under conditions that only the Earth could provide. You've just felt that."**

"Isn't the Earth doing the same thing today?" Katy asked.

"Yes, it is, but you humans are burning it much faster than it is being produced or found. You see, these natural deposits are only found in very special places. Places where they're not too deep or hard to get to."

Ergon went on to say that when the resources are found and identified, they are named "reserves." The problem is that the known reserves have a very short lifetime at the rate people are using them. Oil is the rarest, natural gas is not quite as rare, and coal has the most reserves, but none are going to last much longer unless something changes. And once they are used, they are gone forever.

"Oil has only been heavily used for the last hundred years," Ergon said. **"To the Earth, that's just a blink in time."**

"Why can't they just make a law to stop using it?" Mark asked.

"It's not that simple. Your whole world runs on oil, gas, and coal. Without it, your life would change so much, it wouldn't be at all like your world anymore."

Ergon told Katy and Mark that the industrialized countries of the world—the United States, European countries, and Japan—used most of the world's fossil fuels in transportation, manufacturing, and heating.

"Well, to save the Earth, I could do without it," Mark snorted.

"Oh? Could you? How do you plan to get out to the mall, on a dinosaur?" Katy asked.

"Gosh, I see what you mean," Mark said, although the idea of riding a *Tyrannosaurus rex* sounded good to him.

"Vehicles are a big part of the problem, but oil and gas are used in your society by almost every industry. I want you to see just some of the ways fossil fuels are used so you can tell me how your planet can make them last longer." As Ergon spoke, the group began to move upward. Now they could sense that they were inside a big pipe that ran up to

the surface. Once they were at the surface, they made a short bend and traveled a little distance to a large tank.

In this tank, the oil and gas were separated to go to other storage areas, Ergon explained. As oil, the group went into another large tank for storage, waiting to be moved into a larger pipeline.

"As we begin this journey," Ergon said, **"tell me the steps where damage to the environment is possible."**

In only a moment, they felt themselves pushed out of the large holding tank and into the pipeline. Ergon made it possible for them to see through the wall of the pipe like it was glass, instead of steel.

Soon, they were picking up speed and crossing the broad tundra, or frozen plains, of Alaska in the pipeline.

"Ergon," Katy said, "it's getting warm in here."

"Move in closer to the center," Ergon replied. **"Out by the walls of the pipe the fluid moving against the metal causes heat loss, just like rotating machine parts."**

"So if the oil is warm, I suppose the pipe could melt the ground where they're touching," Katy said.

Ergon said that was correct. The pipeline they were traveling in, the Trans-Alaska line, is about 800 miles long and about half of it is aboveground. In the parts underground, the pipe is well insulated to protect the frozen ground around it. Pipelines all over the world carry energy fluids from points of production to refineries and distribution centers. Ergon told the children that pipelines can even carry coal over long distances if the coal is mixed with water to form a liquid mixture called slurry.

"That must use up a lot of water," Mark said.

"Yes, it does," Ergon answered. **"And in dry parts of the world, like your western United States, many people feel that there are better ways to use water."**

They were passing over a beautiful river, one of more than two hundred rivers and streams the Alaskan pipeline crossed in its long journey from the frozen north shore in the Arctic to the port of Valdez.

Katy brought up the possibility of the pipeline bursting or leaking oil over the beautiful landscape, harming plants and animals.

Ergon agreed that with pipelines, that is always a possibility. This one, he added, is particularly well maintained because it passes through so many wilderness areas. **"A bigger problem may be with the thousands of miles of natural gas pipelines in the lower United States."**

"What do you mean?" Katy asked, with concern in her voice.

"Most of the natural gas in your country is moved by pipelines or transmission lines that were buried in the earth years ago. Over the years, they age and weaken even though the companies that own them usually try to maintain them and keep them safe."

Ergon explained that it is likely that sometimes old pipelines and storage tanks will fail. When that happens, with natural gas or other gas-based oil products, the result is an explosion that could damage people and the landscape for miles around. Some of the gas products, like liquid propane, are unusually dangerous. Liquid propane is heavy and tends to sink down to the ground and not dissolve into the air quickly.

"I heard of a farm family near our town whose well was poisoned by gasoline leaking into the water. The gas was coming from an old service station that was closed years ago," Katy said.

"That is a real problem with underground gas storage tanks," Ergon said. **"The steel in the tanks rusts out and a chemical part of gasoline called benzene gets into the water table. It can cause cancer when people drink the water."**

Mark was watching a good-sized herd of elk in the distance. He had an idea. Couldn't a huge pipe like this stop the animals from crossing to the other side?

"Very good, Mark," Ergon said. Mark was proud of himself for thinking of it. **"When the Alaskan pipeline was designed, the engineers had to make ways for the animals to cross because it lies directly in the path of the migration route for thousands of them."**

Now the group was approaching the port of Valdez. They could see big storage tanks and tankers waiting to load the oil. The group went through the pipeline and into the first tanker they saw. Inside it was like a giant cave, dark with no windows or light coming in from anywhere.

"Is this like the ship that just ran aground and dumped all its oil on the Maine coast?" Katy asked Ergon. He replied that it was similar in size and design.

"Well, isn't there any way to make them safe in an accident?"

Ergon told her that there was no 100 percent guarantee that oil tankers could be made perfectly safe. However, they could be made a lot safer if the shipping companies would build tankers with double hulls. Mark asked Ergon to tell him what a double hull was and what it would do.

"Double hulls are like building a ship inside of another ship," the creature said. **"If the tanker has an accident and the outside hull is ripped, the oil is held safely within another hull on the inside. It's expensive to build ships this way, and that's why it hasn't been done very much. However, it's not as expensive as trying to repair the damage after a big oil spill."**

Ergon told the children that they were going to fly down to the next stop on the journey because if they stayed in the tanker it would be weeks before they got to the next port. With that, they flew out of the ship and headed south. Katy and Mark were both thinking about all the ways that the environ-

ment could be hurt by damage to the pipelines. Spills, leaks, explosions—all were possibilities if the pipelines and oil product carriers weren't constantly kept up and improved.

Ergon picked up their thoughts and said to them, **"Yes, the petroleum industry in the world is large and complicated. Just about every part of your life is touched by it. Even with more people demanding that the industry get environmentally safer, spills and accidents happen every day and never get reported. The public only finds out about the big ones."**

"So how do we stop the accidents that harm the Earth and the wildlife?" Mark wanted to know.

First, Ergon explained, you have to know what's going on. That prepares you to do something about it.

"We certainly are finding out what's going on," Katy mumbled under her breath.

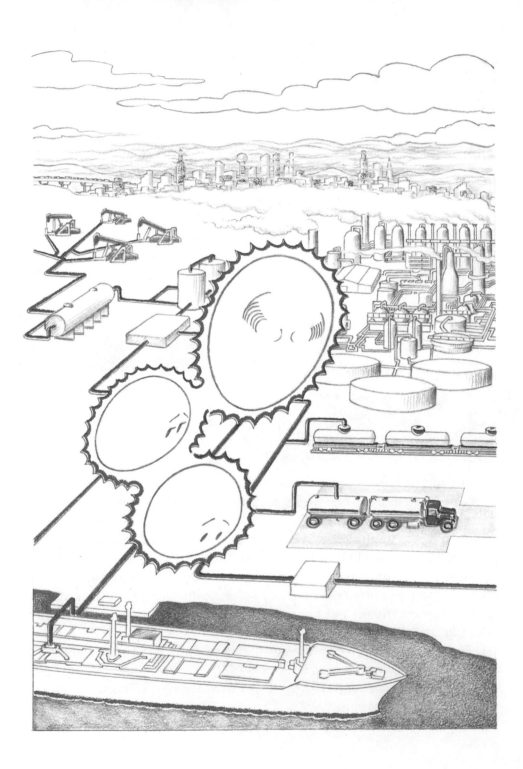

8

ONE PETROBURGER,
HOLD THE MAYO

The group was approaching a great complex of ships, pipes, storage tanks, and buildings on the shore of a large bay. In the water, big tankers and barges were coming in and out of the harbor. Close by, they could see a large, smog-covered city of skyscrapers and freeways. Everywhere cars and trucks were hurrying along the streets and bridges. It seemed that about the same number were going one way as were going the other. They stopped above the vast maze of pipes and buildings and hovered in the air looking down on it.

"This is a refinery complex," Ergon said. **"Here is where the crude oil from all over the world is changed into different kinds of products."**

"What kinds of products?" Katy wanted to know.

"Gasoline, butane, propane, mineral oil, lubricating oil, greases, waxes, fertilizers, plastics and other petrochemicals, to name a few."

"All that from oil, that gooey black stuff we were in?" Mark asked.

Ergon explained how the crude oil is heated by high temperatures and separated in the tall skinny structures below. The heating breaks up the crude oil into light, medium, or heavy substances that are then treated to produce raw materials for making other products. The heavier products are ones like asphalt, used to pave roads and make playgrounds. The lighter and more complex ones are jet fuel and valuable chemicals. Of course, Ergon said, all these conversions require heat energy, and some energy is lost in the process.

"It's another example of energy changing form," Ergon added.

Mark had been thinking about all the things in his home made of plastics. Parts of the furniture, countertops in the kitchen, picture frames, the case of his stereo, cassettes, most of his toys—the list was endless.

"I've never thought of all those 'things' as energy before," he said to Ergon. "But they are!"

"You see what I meant when I said your world owes its life to plants and animals from millions of years ago, don't you?" Ergon said to Mark.

Ergon continued. Not only are the materials themselves from oil energy but the processes used to make and deliver them all come from petroleum. Below them on one side of the refinery complex Katy and Mark could see trucks, railroad cars, and boats loading up.

"Are those trucks taking the oil products to different manufacturers?" Katy asked. Those, as well as the trains, boats, and pipelines leaving the refining center, Ergon answered. Some of the products like gasoline are ready to use, but most must go through many more conversion steps to become finished products.

Ergon took the group out of its hovering pattern over the refinery, and they began to follow one of the train tracks below. Soon they came to another cluster of immense buildings where the train cars lined up alongside one structure. Here, Ergon explained, the oil-based materials from the refinery are used to make resins for different kinds of plastics.

"At this point, a lot of dangerous things can happen to disturb the environment," Ergon said. The making of plastic resins combines different oil-based molecules to form new complex chains of molecules. Some of the manufacturing methods that have been used for years were found to be harmful to the atmosphere. Also, the solid and liquid wastes from the manufacturing process are often toxic.

"Oooogh," Mark said with excitement. "I saw the old movie *Toxic Avenger.* All that gunk gets on your body and changes you to an ugly monster! It was awesome."

"That's a bit of a story, Mark," Ergon said. **"But wastes from manufacturing that go into the air, water, or earth often do great damage to the environment."**

"Why don't you zap it, Ergon?" Mark yelled. "That would teach them not to pollute. Just hit that smokestack with one of your energy beams, and BAM, no more problem." Mark was getting really worked up. As he was yelling ideas to Ergon, he was diving down toward the plant making noises like electronic rifles and pretending he was a missile aimed at the factory.

"HOLD IT, MARK!" Ergon's voice split through the air. Mark quickly pulled up and away from his dive bombing and came back to Katy and Ergon.

"I'm sorry, I got a little carried away again," he said quietly.

"It makes no sense to destroy something because you don't like the way it operates. Instead, you should work to change it." Ergon addressed the statement to Mark, but Katy answered.

"Ergon, it's a little frustrating when you're a kid to see all these huge energy factories that your parents or their parents built. How can we change it? The damage is done. There's nothing we can do." Katy's voice was firm as she looked Ergon directly in the eyes.

"THAT'S WHERE YOU'RE WRONG!" Ergon said. He gathered Katy and Mark to him as they shot into the upper atmosphere. The group was headed back to the eastern part of the country. Across the continent they flew. Below, as they sped into darkness, they could see bright groups of lights that were the large cities. In many areas, the interstate highway system connected the cities to each other like a dew-covered cobweb in the early morning. Now the children could sense they were headed down. The lights of their town grew brighter. Then they were at treetop level speeding above streets and buildings they recognized.

When they passed above the familiar mall surrounded by fast food restaurants, Mark suddenly veered off and down toward one of the hamburger stops.

"I'm starved," he shouted. "I never did get my afternoon snack." Ergon and Katy followed him down.

Katy shouted back at him. "Mark, you're in energy form! You don't even have a body right now." But as she was trying to get his attention, she was shocked to see him turn back into his human form and walk through the swinging door.

Ergon said, **"I think there's a lesson here for your brother. Let's go meet him as he comes out."** The two of them landed in a dark spot alongside the restaurant. Soon Mark came through the front door carrying a large paper sack and wearing a big smile.

"We're over here," Katy said, catching his attention.

Mark could see his sister and the energy being by the faint glow they were making in the darkness. "You ought to join me, Katy," he said, sitting down on the curb. "I have a little more money left. I'll loan you some if you pay me back tomorrow."

Katy said no thanks, she wasn't hungry and watched Mark as he unloaded his

goodies. First out of the bag came a large paper cup with a clear plastic top. Next out was a styrofoam box, followed by five folded paper napkins Mark had caught between his fingers. Then he reached back into the bag and pulled out a cardboard holder stuffed with French fries. Finally, looking into the sack and fumbling around, Mark grabbed four little ketchup packages, two plastic straws covered with paper wrapping, and five or six tiny salt and pepper containers. He laid all of this out around him as he sat on the curb.

"You know, Ergon," Mark began talking with his mouth full of fries, "Katy's right. It really is hard for a kid to do anything about saving natural resources and energy." He put down the fries and pried open the styrofoam box, revealing the double cheeseburger inside. "Brrr. It's cold out here if you're a human. I better slam dunk this stuff in a hurry before it turns to ice."

Ergon said nothing but watched as Mark got his hands around the burger and pushed it toward his mouth. Suddenly, a flash of light shot out from Ergon, and with a loud **Crackkkk** the cheeseburger had disappeared. In its place was a thin wisp of smoke drifting up from Mark's fingertips.

"Aw, what did you do that for?" Mark whined. "That burger was *my* energy source."

"I'll get it back in a moment, if you wish," Ergon said.

"I wish! I wish!" Mark mumbled, but Ergon kept on talking as if he hadn't heard the boy. Meanwhile, Katy noticed all the containers had been rearranged by Ergon. The paper bag, napkins, drink cup, and cardboard French fry box

were in one bunch. The styrofoam box, the lid for the soft drink, the ketchup packages, and the straws were in another neat pile.

Ergon was speaking. **"Let's look at all the packing materials here, and you tell me what resources were used to get them made and to get them here."**

Katy looked at the groupings. The paper goods were the most obvious, so she said, "Paper, that's made from trees."

"Renewable or nonrenewable?"

"Renewable, but," she paused, "I think there's a problem with overcutting our trees and using them faster than new ones are planted."

"Correct." Ergon asked Mark the next question.

"Is there any oil used to deliver these products?"

Mark was still angry at Ergon, but he said, "I think so. After all, they had to be delivered here." Mark thought some more. "Also, they've got all kinds of colorful ink on them, and that might be oil based. And, sure, the products had to be folded and packed by machines. That probably took oil and electricity."

"Very good, Mark," Ergon said. **"Now, can these products be used again?"**

"Sure, they can be recycled as new paper products," Mark answered brightly, forgetting his vanished burger. "That would save trees and energy!"

"How about this group?" Ergon pointed toward the styrofoam box, plastic drink lid, straws, and ketchup packages.

Katy said she thought that group was almost pure petrochemicals. The raw material comes from oil, the forming, coloring, and packaging is oil based, and the shipping is by truck. In addition, Katy said, producing the styrofoam created by-products in the atmosphere which harms the ozone layer around the Earth. If you just throw styrofoam away it lasts almost forever and doesn't turn back into anything the Earth can use. She went on to say that the plastic products could probably be recycled, but she understood that some plastics were easier to recycle than others.

"So, if the paper and plastic can be recycled, where are the recycling containers?"

The children looked around and didn't see any.

"Maybe a better question is, Why do you need so much of this packaging stuff in the first place?" Ergon asked. He went on to suggest some things that Katy and Mark should think about. Like, why couldn't the burger be wrapped in paper? Did the paper cup really need a plastic lid? If you needed one, you could ask for it. Same with the napkins. Why five of them if one would do? And who needed four ketchups and six salts for one small order of fries and a burger, anyway?

"So what's a kid supposed to do about it?" Katy asked Ergon.

"Pick up all this oil and tree matter, take it back into the burger place, and ask the manager where the recycling bins are. If there aren't any, ask WHY NOT. Then, Mark, get your food without the bag, styrofoam box, or lids."

"They'll think I'm weird," Mark protested.

"No, they'll think you're a kid customer who, along with other kids, makes up over half their business, and they'll do what you want."

Mark did what Ergon asked and came out in a few minutes smiling. He sat down in the same place with a burger wrapped in paper and a soft drink without a lid. One straw was sticking up out of the side of the cup.

"You were right, Ergon. They didn't hassle me at all," Mark said. "By the way, you owe me two and a half dollars for the food you vaporized." He carefully unwrapped the paper from the cheeseburger and was getting ready to bite into it when it floated out of his hands and up over his head. It hovered there like a flying saucer.

"Hey," he shouted. "What's going on now? Can't I ever get something to eat? A guy could starve to death hanging around with you, Ergon. Gimme that burger!" He stood up quickly and made a jump for it, but the cheeseburger just drifted up a little higher out of his reach.

"There's one more thing I want you both to know about how energy is used in food production and how that can affect your environment," Ergon said. As he spoke and Mark continued to jump for the cheeseburger, the food slowly changed its appearance. Instead of a juicy burger, it was the Image of a cow grazing in a meadow.

"I prefer my meat cooked," Mark said to Ergon. But he sat down on the curb and stared at the image of the cow in wonder.

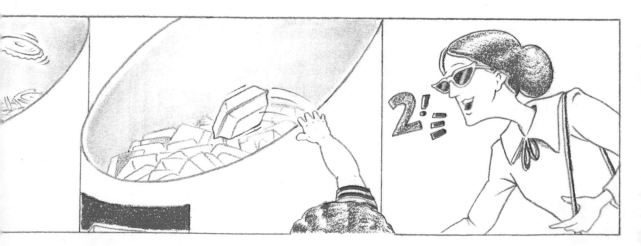

Ergon began to explain. **"I'm not going to try to tell you how to eat, but I want you to think of the energy that goes into your food so you don't take it for granted."**

Katy and Mark should think of all the plants and animals of Earth linked together in giant chains, Ergon said. At the top would be humans. A cow would be closer to the top but below wolves and sharks. At the bottom would be the microorganisms of the Earth and seas. The food chains show the link between organisms that eat or live off of other things lower on the food chain or get consumed by things higher. Carnivores, the meat eaters, and omnivores, who eat anything, are at the top.

All creatures depend on the very lowest at the bottom of the chain in the sea or soil. Just like anything else, food is an energy conversion process, Ergon explained, with the waste energy going to heat the environment. For instance, grass converts sunlight to carbohydrates for the body of the plant. But only about 1 percent of the sun's energy striking the grass is converted. A cow, eating the grass, can only convert about 10 percent of the energy in the grass to energy for its body. A person, eating the cow, can only get about 10 percent of the energy from the meat into useful form for the human body.

Katy thought about this a while. It seemed that humans, the animals at the top of the food chain, are very inefficient eaters. They don't use most of the

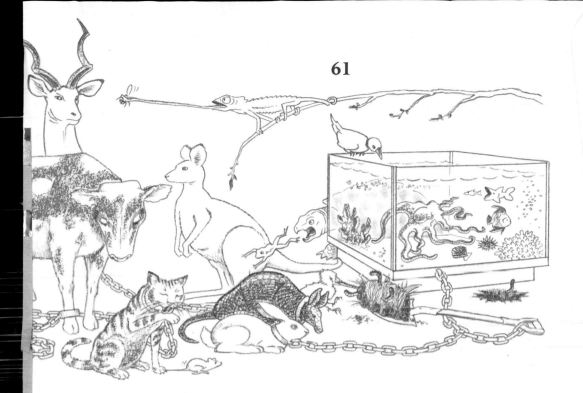

energy in the animals they consume. "So, does that mean that the most efficient use of food energy for humans would be to eat lower on the food chain?"

"You've got it, exactly!" Ergon said, glowing a pleasing orange. Of course, he went on to explain, people are going to eat what they like, but they should understand how little of the actual energy they need to live is provided by animals. Most of their energy needs can be met by fruits, vegetables, and grains.

"To add to the energy problem, much of your food is highly transported and processed, needing petrochemicals every step of the way. That processed cheeseburger is a great example. One way to save is to grow more of your own food. When you grow vegetables in a home garden or greenhouse, you not only get wonderful-tasting food but you actually save oil. The garden tomato or cucumber doesn't have to be harvested by machine, cooled, stored, canned or boxed, packaged, and shipped a thousand miles."

The image of the cow was still floating above Mark.

"Enjoy your Petroburger," Katy said to Mark.

The burger floated gently down into Mark's hands. He took a long look at it and then glanced at Katy. "Later," he said.

Instant Energy ➔ Energy used <u>once</u> to do a temporary job.

Process Energy ➔ Energy used to make something.

Long-Term Energy ➔ Energy contained in an obje

9

THE OIL WELL AT YOUR HOUSE

Ergon reassembled the group in Katy and Mark's living room. Mark was standing by the TV, which was off, with a cold cheeseburger in his hand. He looked at the clock and noticed that only about an hour had gone by since the magical energy being first made his surprise appearance.

"How come so little time has gone by while we've visited several other worlds and a lot of this one?" Mark asked Ergon.

"When you're with me, time only passes when you are in human form. We were traveling in energy form a great deal."

But Katy was thinking about the big problems with energy and the environment they had discovered. Her brain felt overloaded and tired, but she wanted some answers. She wanted some things to do that would make a difference!

"OK, Ergon," she said. "Now I think I understand something about how energy is converted, transported, used, and misused. What do I do about it? The problems seem so big, and we don't have all the time in the world, you know. We're human, again."

"You start by seeing everything around you as some form of energy. I think you now can do that." The children agreed that they had learned to see objects and actions as energy.

"I'd like you to think of energy by the way its used: instant, process, or long-term."

In the air in front of them, Ergon created a big sign:

Instant Energy ⇒	*Energy used once to do a temporary job*
Process Energy ⇒	*Energy used to make something*
Long-Term Energy ⇒	*The energy contained in an object*

"I think I see what you mean, Ergon." Mark said. "It's mostly the length of time energy stays in one form to do the work."

"Yes," the being answered. **"Let's look around your house and see the different ways energy is used in products."**

"Let's start with the TV!" Mark said.

Katy began. "Let's see. The case, all the dials, and the trim look like plastic. They represent *long-term energy* because the pieces will last for years. The TV tube itself is glass, and some metal and plastics, I guess. It must be long-term energy also."

Mark spoke up. "And to manufacture the parts, inside and outside the TV set, took the *process use* of energy. So the parts themselves are long-term, but the manufacturing it took to make them represents process energy. Is that what you mean?"

"Exactly. Long-term energy is locked in products and is released if the product changes form."

"When we turn the TV on, then we're using electricity from the grid for the *instant use* of energy, right?"

"So which form of energy can be used and reused over and over again?" Ergon asked.

Both kids were stumped. After a few moments, Katy spoke.

"I suppose many of the parts of an old TV could be reused," she said. Now she was starting to see what Ergon was saying, and she began to talk faster. "And if the parts are reused and put into a working TV, then new parts wouldn't have to be made! So no petrochemicals would be used for the parts themselves or the process of making them."

"It's kind of like a mine or an oil well in your living room," Mark said. He was beginning to understand what Ergon was driving at. "All these objects we have are energy forms that can be used and reused. If that were done, we'd save both the energy in them and the waste they create at the dump."

"What about the instant energy used for powering the television and other appliances in the house?"

"It's gone as we use it," Katy replied.

Ergon said that was right; instant energy is history the moment it does the work. When products are made, a large portion of the energy *in them* can be saved *if* the products are reused or recycled. The process energy used to make the products, like instant energy, is gone into the atmosphere forever.

Next, the group went into the kitchen. Opening the door to the water heater, Ergon pointed at it and said, **"Earlier I told you that this was one** of the bigger energy wasters in your house. **Now you tell me why.**"

Both children stared thoughtfully at the old water heater. It was easy now for Katy and Mark to see the long-term use of energy contained in the steel of the tank and the pipes and metal vent leading out of it. They could feel the heat coming out of the closet from the machine. Mark spoke, "It's making heat, but it's not doing anything. I mean, we don't have the hot water turned on anywhere in the house."

"So, the heater is using instant energy—gas, I think—to heat water all the time, but we only use the water every once in a while," Katy said. She was thinking about all the oil and gas being taken out of the ground, pumped, refined, and transported into her house so this tank could make heat that wasn't being used most of the time. "This is really wasteful," she concluded.

Instant energy use is necessary for many jobs, but it is often wasteful and based on machines and practices that are fifty years old.

"You see," Ergon said, **"there's smart energy use and stupid energy use. Smart use is finding the proper energy source for the job and seeing that the smallest amount of energy is used to do the job right. Stupid energy use is not paying attention to how much energy is needed to do a job. It's like stuffing yourself with a huge meal before a baseball game and then wondering why you can't chase fly balls."**

Ergon went on to say that when oil and electricity were put into wide use about a hundred years ago, people thought that the natural resources—oil, coal, gas—could never run out. When early water heaters, furnaces, air-conditioners, and automobiles were made, no one thought about how efficiently the machines used the energy. Customers wanted machines that ran without problems, but how much energy they used was not something they worried about. Also, because the skies were clear and the land uncrowded, people felt that waste from the inefficient machines could always be dumped somewhere else, into the land, water, or air. So, Katy and Mark's generation inherited a vast amount of power in machines, but these machines were based on old-fashioned energy ideas, inefficient and wasteful in their design.

He led the children out of the kitchen and into the study. Ergon pointed to the small computer on the desk, and it flashed to life. The whole family used the little machine: mom and dad for business, the kids for their schoolwork and games.

"Can you think of anything else that would power this machine besides electricity?" Ergon asked. The children looked at the glowing display tube and thought about it. They had learned many ways that electricity could be made *for* the computer, but they couldn't think of any way the computer, screen, and printer could run without electricity.

"Now put your hands on top of it," Ergon directed. They did and felt a very slight warmth coming from the case. They knew it was a small amount of waste heat being lost by the parts inside the machine.

"So, would you say that this was smart or stupid use of instant energy?" he asked.

"Smart!" both kids said at once.

"Smart in three ways," Ergon explained, turning off the computer. First, the machine is only on when someone wants to work, unlike the water heater, which works all the time to make hot water people use only occasionally. Second, because the electricity is used to run circuits and microchips, very little energy is wasted in heat. Third, properly cared for, the little computer would run for years. If the family wanted more power or speed from their computer, new parts could be added.

"In fact," Ergon finished, **"you can probably run this computer for ten years on the amount of energy the water heater wastes in a week. Now let's see another side of energy use."**

Ergon took Katy and Mark down to the basement and into the furnace room. Dad had told them not to go into the room unless he or mom was there, too, but with Ergon along, Katy thought it would be all right. When they opened the door, the kids felt a blast of heat in their faces. Inside the room filled with pipes, machines, and wires was a low rumbling noise. Ergon explained that the noise was made by the fans pumping heat from the furnace into different parts of the house.

Mark could see that some process energy was certainly used to manufacture the parts of the heating system, but it looked to him like a lot more energy was going into keeping it running.

"This is kind of like the water heater," he said. "All the machinery is going

like crazy to keep the whole house warm when we're just in a part of it."

"OK, Ergon," Mark said, "I'm beginning to see how there is smart energy and stupid energy, but how do you make this stupid machine smart?"

"For starters, make sure your parents have it checked and tuned once a year so it gets the cleanest burning it can."

The group then went upstairs, and Ergon showed them a device on the wall called a "thermostat." This little machine contained a thermometer and sensor to tell the furnace when to send heat to different parts of the house. Going around the house, they found four thermostats. Three were set at 78 degrees, even those in the part of the house where the family slept. The one in the hallway by the living room was set at 85 degrees.

Ergon asked, **"Were you using the bedrooms, the study, and the kitchen when I arrived?"**

Katy and Mark said no, although they had gone into the kitchen a couple of times to fix food.

"Then you could turn down three of the thermostats in parts of the house you're not living in, couldn't you?" the creature continued. **"Furthermore, if you put on sweaters, you could be comfortable in the living room at 70 degrees."**

The children agreed that was true.

"The best way to make this furnace of yours smart is to keep it turned off when it isn't needed," Ergon said.

"Maybe," Mark said, "but we don't want to freeze to death in the dark."

"THERE'S NO REASON FOR THAT!" Ergon replied loudly. His color was changing again.

Mark shrunk back into the corner, but Katy stepped up to defend her brother. Standing up to the large mass of pulsing red energy, she looked small, but her voice came out clearly and firmly.

"Why are you yelling at him, you big bully. He's just afraid that all this energy saving you're talking about will make us cold. Why don't you learn to control your temper?" She was not backing away an inch.

Ergon seemed to calm down as his color lightened again. The creature paused for a few moments and said, **"Mark, I apologize. Your sister is right. There was no reason for me to yell at you."** He went on to explain that

the fear Mark expressed, freezing in the dark, is one of the biggest mistakes people make about conserving energy. Many people think energy conservation means giving things up, being uncomfortable or crowded, made to do without. Actually, conservation is simply finding better ways to get the job done with a minimum of waste and no change of comfort. **"In many ways,"** Ergon said, **"conservation can make your life more comfortable."**

For instance, one way to make the furnace smarter is a computer-controlled thermostat that is programmed for the way members of the family live. It can be set so that heat in the bedrooms and bathrooms goes on at 6:00 a.m. when people are getting up and dressing. It can turn off in those rooms at 7:00 a.m. and turn on the heat in the kitchen area when the family is having breakfast. When everyone leaves for work and school, the smart thermostat can turn down the home's heat until around 3:00 in the afternoon when the kids return from school. Then it can turn on heat in the living room while they study and watch TV. By working in this way throughout a full day, the machine can save one-fourth of the energy the home is using. The device can even be set differently for Saturdays and Sundays when the family is home.

"Now that's a smart machine!" Mark said.

"Your whole house could be a much smarter machine," Ergon said. **"The biggest problem is that a lot of the heat energy is leaking out of it."**

"How does that happen?" Katy wanted to know. In her mind, she saw a boat with lots of holes and water pouring into it.

"Do you want to be molecules again for a few moments?" the creature asked. **"It's the best way to show you."** Katy was surprised, for this was the first time Ergon had ever stopped to ask them if they wanted to change.

"Sure," said Katy.

"I can handle it," said Mark.

They were now floating with Ergon along the carpet on the floor of the living room. **"So that you can see how heat moves, I'm going to give you two qualities that real air doesn't have. First, you'll be able to move through solids, and second, you'll have a sense of temperature."**

"Sounds like fun," Mark yelled, as he started rising higher in the room.

"Right now you're rising because you're lighter in weight than the cooler air around you. This movement of heat through air, or any gas or liquid, caused by the motion of the material is called *convection*," Ergon said.

Katy and Mark were bouncing upward on the air currents and stopped when they hit paneling above them. **"Up here, at the ceiling, it's usually 10 or 15 degrees warmer than at the floor."**

Katy could feel that she'd gotten warmer and was bouncing around more the higher they went.

"When heat moves through any material by means of a warm molecule becoming excited and passing heat to the next cooler molecule, it's called *conduction*. **Are you excited and warm enough to go through the ceiling?"** Ergon asked.

"I'm warmer than the surface of the ceiling and definitely more excited," Katy said with a laugh.

"Then here we go!"

With that, the group entered the dense paneling of the ceiling and passed through it into a pink cotton candy world of swirls and ribbons. It was much cooler here.

"What's this stuff we're in, Ergon?" Mark asked.

"This is insulation in your attic," Ergon replied. **"It helps keep your house warmer in winter and cooler in summer."** They started moving up through the pink material, and Ergon explained as they went. **"These small air pockets in the insulation are what keep the heat from escaping. Each one acts as a trap and slows the movement of heat through the material."**

Katy and Mark could feel that most air movement, the bumping around they felt in the living room, had stopped; it was almost perfectly still. The inside of the insulation was like a million small caves, each one separate but connected to the next one.

When the three got to the top, they passed through a thick paper and foil barrier. It was cold above the insulation in the unheated attic.

"You've done just what heat always does, move to a colder area." Ergon was pointing to the floor. **"Hmm, there's room up here for at least four more inches of insulation."** Over the years, the insulation in Katy and Mark's attic had packed down. Now there was enough space to add more. **"Over half of the heat loss in a house is through the roof. It almost always makes sense and saves energy to add insulation,"** Ergon said.

The group was being blown around quite a bit by winds coming through some cracks between the roof and the walls. **"Here's a big energy waster."** Ergon showed the kids the open areas to the outside. **"Careful, don't get blown out of here."**

The group continued floating above the floor of the attic insulation until they

got to the large pipes coming up through the ceiling. The pipes continued through the attic and went out of the roof above. Looking below where the metal came through the ceiling, Katy and Mark could see all the way down into the house. Ergon told them that the openings around vents, pipes, and chimneys should be sealed with a special caulk to keep cold air from falling through the cracks into the living areas of the house.

The cold air currents are strong around the vent openings, Ergon said, as he jumped into the crack between the ceiling and the pipe. Katy and Mark followed and found themselves tumbling down into the hot water heater closet.

"The best way to make this machine smarter is to put a jacket on it," Ergon said. He told the children that for a small amount of money, they could get an insulated jacket to wrap up the water heater. The insulation in the jacket keeps more heat in the water and lets less waste heat escape into the air. One poorly insulated electric water heater is responsible for 10,000 pounds of CO_2 released into the air at a coal-fired electrical power plant.

"After you do that, see about putting water saving shower heads and flow restrictors on all the showers and faucets in your house. Those simple and inexpensive devices reduce the amount of heat and water used and the amount of pollution released into the air." Ergon gave the example that each low-flow shower head installed would keep over 1,500 pounds of CO_2 from entering the atmosphere by restricting the amount of water used and therefore the amount needed to be heated. As the creature spoke, he was leading Katy and Mark back into the living room. This time the group was drifting toward the top of a large picture window.

"Now you're really going to feel some convection," Ergon said. Katy and Mark were still bouncing along the ceiling as warm air when they approached the top of the window. As both got close to the glass, they were suddenly caught in a severe down draft. It was like stepping off a cliff, as the two molecules fell toward the floor at high speed. The glass rushed by; the sill of the window rushed up to meet them.

"What a ride!" Mark yelled at Katy, as he narrowly missed the window sill and continued down.

Ergon was waiting near the floor as Katy and Mark leveled out. **"You've just felt what it's like to be a warm air molecule that gets close to a cold glass surface,"** Ergon said.

"Is this going on at *all* our windows?" Katy asked. Ergon replied that, sadly, it was. Katy's house had old single-pane glass windows and the conduction of heat right through the glass was very rapid. This created a cold surface on the inside of the glass where it met the warm air of the room. When that happened, it rapidly cooled the inside air and caused it to fall. Anyone sitting close to the windows in winter would be very cold and uncomfortable. The result: cold people, an uncomfortable house, and wasted energy as the furnace kept trying to warm up the air.

"What can we do about it?" Katy asked.

First, Ergon said, all the solutions will cost some money. Some are inexpensive, others aren't. The cheapest way to prevent loss of heat through the windows is to attach clear plastic on the inside; the kids could help put it up. There are many kinds of products available, and some even include little tracks to help mount the material.

"But wait a minute, Ergon," Mark said excitedly. "You've already shown us that plastic is a petrochemical. Making it takes oil and puts some waste into the air."

"Ah...you're right, Mark, but you've brought up a good example of process energy use versus instant energy."

Mark had a questioning look as the being continued. **"Yes, it does take some energy to make the plastic, but the amount in it is small and well spent compared to the heat that is going out of your windows everyday. Do you know that as much energy leaks out of American windows every year as flows through the Alaskan pipeline?"**

"Isn't there an answer that will save the heat the windows are losing without using any extra fossil fuels?" Mark asked.

"Unfortunately, not at this time, not without turning the heat way down," Ergon said. **"If you take care of the plastic on the window, it should last four or five years. Then the material can be recycled."**

Ergon went on to explain that there are other materials, insulated glass and insulated window coverings, that can do an even better job, but they are more expensive. The best way to convince their parents to do something is to have them put their faces right next to the glass on a cold evening. Ergon was sure that would convince them to spend some money on the problem.

"Let me show you one more thing about cold surfaces and energy," Ergon said. He took Katy and Mark to the other side of the room and changed them back to human form. **"Put your hand up to your cheeks and tell me what you feel."**

Both children did. Katy spoke. "It's warm. Maybe 98.6 degrees. That's body temperature, isn't it?"

"That's internal human body temperature. At the surface of your skin, the temperature is around 90 degrees. Now go feel the glass in the window."

Katy and Mark did so and told Ergon that it was very cold. Ergon replied that it was below freezing outside and so the single glass would probably have an inside temperature in the low 40s. There was about a 50 degree temperature

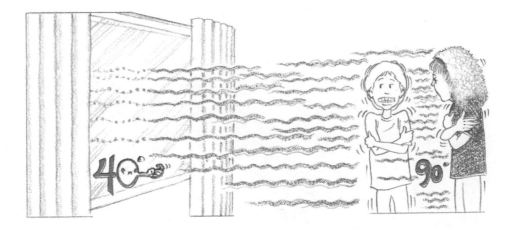

difference between their skin and the window glass. **"What does heat have to do?"** Ergon asked.

"Move to cold!" Katy and Mark yelled together.

"And so it will," Ergon said, looking pleased. **"Across a room or across a solar system, heat will go directly to cold by *radiant* transfer of electromagnetic waves through space. This heat loss from your skin to cold glass surfaces will make you chilly even if the air temperature in a room is hot."** He ended by saying, **"Making your windows save heat is a great example of how conserving energy can make people more comfortable."**

A moment after Ergon finished speaking, all of the lights in the house came on, exactly as they had been when the energy creature first appeared to Katy and Mark. Both children looked at each other in embarrassment. Now, they both knew what it took in oil, gas, and coal to keep all these lights working. Katy ran toward the kitchen and Mark headed for the bedrooms to turn lights off. Ergon let them run around the house for several minutes doing the job.

"That's the first way to save the wasted lighting energy," he said. **"Turn off lights that aren't needed."** They were left with one small light in the living room, which was perfectly adequate.

Ergon took the kids over to the lamp. He told them that this *incandescent* light bulb, along with the billions like it in the United States, consume about 25 percent, or one-fourth, of all the electricity made in the country. In recent

years, new technology had created compact *fluorescent* light bulbs that need only a small fraction of the power to produce about five times the light. One single new light of this type will keep a ton of CO_2 from entering the atmosphere and will last over 10 times as long as the old type of bulb. This was another example of conservation making life better by saving energy while it saved the hassles of buying and replacing light bulbs!

Ergon said he wanted the children to see something else. He dissolved the group in the house and reassembled them on the driveway.

Katy and Mark weren't wearing jackets and the cold wind felt as if it was blowing right through them. Ergon told them to move closer to him. His body was a warm glow radiating heat as Katy and Mark moved next to him.

"Watch the cars," Ergon said.

On the street in front of the house a constant string of cars was passing by as people hurried home from a long day's work. Ergon told Katy and Mark he wanted them to count how many people were in each car. It didn't take long for the children to discover that almost every car had only one person in it. Many of the vehicles were big ones—vans, trucks, and the four-wheel-drive wagons—that had become so popular in the last couple of years.

"Here you have a major form of energy waste in your country," Ergon said. **"Not only did a great amount of process energy and long-term energy go into making the vehicles but the big waste, the waste of instant energy to move these big hunks of steel around with only one person in them, is shameful."**

Ergon explained to the children that it takes hardly any more gasoline to move four or five people in a car than it takes to move one person around. If one gallon of gas can take a car twenty-five miles and there is only one person in the car, then one gallon equals twenty-five passenger miles. If you put five people in the same car, the car might only get twenty miles to a gallon of gas because the load is heavier, but that one gallon will have moved *one hundred passenger miles*. The one gallon of gasoline has done four times the work.

In a way, he explained, it is like making extra gallons of gas, only better, because no instant or process energy goes into creating or distributing it. Public transportation—buses, subways, and van pools—are even better because more people can be moved for the same amount of energy.

"We both ride buses to school," Mark said. "And they put about fifty kids on the bus. It's crowded, but it's a lot of fun."

"You can find more ways your family and friends can travel together to get where you want to go," Ergon said. He went on to explain to the kids that vehicles are one of the greatest sources of various types of air pollution. They account for about one-third of all the carbon dioxide emissions that are a big part of the greenhouse warming problem and one-third of the nitrous oxide in the air that is a contributor to acid rain. In addition, the coolant in air-conditioned vehicles is a chlorofluorocarbon (CFC) that is destroying the ozone layer in the upper atmosphere.

"There are so many ways to save energy and help the environment by improving car use that you can start just about anywhere," Ergon said. On the driveway, he created a list:

1) Don't make, or ask your parents to make, unnecessary trips.

2) Talk to your family and friends about carpooling and sharing rides.

3) Learn to check the tires for proper pressure to get better gas mileage.

4) Remind your parents that a good tune-up makes a car run better and last longer. Every mile per gallon increase in gas mileage means less harmful pollutants in the air.

5) When your parents look at new cars, or you do in a few years, get a car with great gas mileage. Some of the new cars get over 50 miles to the gallon, but the average car that people buy gets only half of that.

For instance, Ergon explained, over its lifetime a car that gets 50 miles to the gallon will prevent about 30 tons, *that is, 60,000 pounds*, of CO_2 from entering the air.

"Can we yell at people who are riding alone in a car?" Mark wanted to know. Ergon said that probably wasn't a good idea, but they could question people they know well about their driving habits. The energy creature could see that the children were shivering as they stood huddled next to him.

"Do you wish to go inside?"

"I'd like to," Katy answered. Then she added with a smile, "Your radiant energy is great, but the convection currents are freezing me!"

With a laugh, Ergon took them back into the living room.

"I've shown you things that you can do with your home and car, or help get your parents to do, for important reasons. First, over half of all the energy produced in your country goes into the home and automobiles you use every day. Because we're talking about such a huge amount of energy, anything you save has a quick and helpful effect on the environment."

"Yeah," said Mark, "some of it is stuff we can do right away."

"Next, you have some control over your immediate environment and the way you live. As you grow up, you will have more control." Ergon paused before he created a sign in the air in front, and slightly above, Katy and Mark.

He pointed to the letters floating in the air and said, **"Here are guidelines you should follow for sensible energy practice in your daily life. They are listed in the order of their importance to long-term energy and environmental savings."**

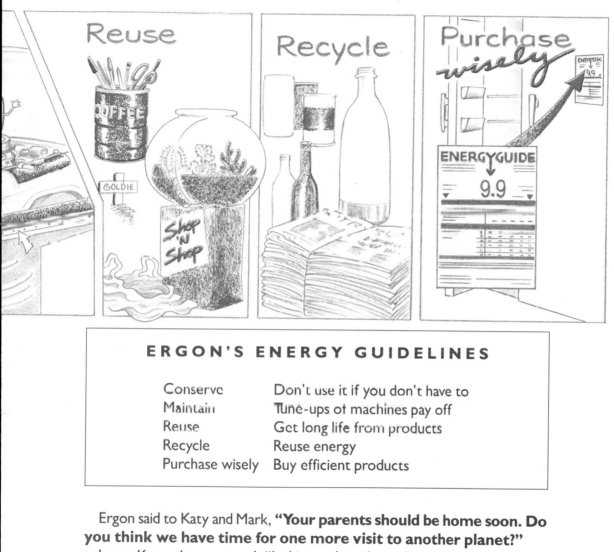

ERGON'S ENERGY GUIDELINES

Conserve	Don't use it if you don't have to
Maintain	Tune-ups of machines pay off
Reuse	Get long life from products
Recycle	Reuse energy
Purchase wisely	Buy efficient products

Ergon said to Katy and Mark, **"Your parents should be home soon. Do you think we have time for one more visit to another planet?"**

It was Katy who answered, "Is this another place where the people messed everything up?"

"If it's that way, I don't want to go," Mark said. "It's just too sad."

"No," Ergon answered. **"In this place, they have done things right."**

10

VISIT TO A SMART PLANET

I t felt good to be out among the stars again, Katy thought, as the group streaked through the cosmos. In the millions of stars around them, she saw the creative energy that made and sustained planets. She hoped that the planet Ergon was leading them to was similar to Earth, enough the same so that she could make comparisons.

A planet directly ahead of them was growing. The round ball was mostly blue, with landmasses of green and tan mixed. To Katy, it looked almost identical to Earth, with perhaps a little more land and less water. She could see lines of white clouds in regular patterns marching across the oceans of the planet.

"Is that it, is that the one we're going to?" Katy aimed the happy thought at Ergon.

"Yes, it is. The name of this planet is Verdant. In so many ways it is like your Earth. I thought you should see what your planet can become." Ergon guided them down toward the globe slowly, as if he wasn't in a great hurry to land.

"I want to tell you a little about the history of this planet." Both children were enjoying the view. Mark answered, "Well, I've missed my snack and dinner won't be ready for a while. I'm in no hurry."

"Not long ago," Ergon began, **"Verdant was in almost the same shape that Earth is today. Her people depended almost entirely on fossil fuels, she'd gone down the atomic energy road and was learning that it was going nowhere, plus the population was growing at an alarming rate. The planet had problems; the resources to save her were being used up faster than anyone could imagine."**

"Sounds familiar," Katy said with some anger.

"In addition, the wealth of the planet was distributed unequally," Ergon continued. **"That is, a small group of nations controlled and used most of the natural resources and had the highest technology on the planet. Of course, the majority of the people who didn't have the products and services wanted everything that the wealthy nations had."**

"Why didn't the wealthy nations just sell it to them?" Mark asked.

"They would have if the poorer nations could have paid. But the wealthy nations had problems of their own. The air and water in them was crummy and getting worse daily. The rich nations were paying so much for the oil to keep them running that they didn't have much extra to give, anyway."

"So what happened to change the situation," Katy asked.

"A turnabout took place; slowly at first, but then it gathered speed and momentum." Ergon went on to tell Katy and Mark how it occurred. **"Citizens, mostly from the rich nations, began to understand that all the goods and services they enjoyed were linked to the well-being of the planet—its water, air, and land. Once the people understood this, they began to form groups of their own and act to change things."**

"What did they do, stop buying things?" Mark asked.

"No, they realized the only thing they could really buy with all their money, energy, and technology was time," Ergon said.

"Time! Nobody can buy time!" Katy answered. "And another thing, Ergon, you said a moment ago that the citizens of the *rich nations* started acting. Do you mean that people in the poorer nations didn't care about the environment?"

"No. I mean that when you're trying to figure out where the next meal for your five starving children is coming from, the sulfur content of the air around you is not your main concern," the creature answered.

"Oh, I see what you mean," Katy said. "The people in the wealthy nations could afford to spend time thinking and acting on the problems. Well, I guess that's only right as they were using most of the energy, anyway."

"So, back to the time thing," Mark said. "How did they buy time?"

"The rich nations discovered that in conservation, reuse of goods, recycling, and smart energy use they had found plenty of energy to meet their demands. Citizen groups and individuals began demanding sensible environmental policies from the national leaders. At this point, kids became a very important part of the solutions."

"Kids? Why kids? That could never happen on Earth," Katy said.

"Oh, no? It already has. How about the teenage girl several years ago who got directly in touch with the leader of the Soviet Union to talk about world peace? Or the boy who convinced the chairman of a large tuna company to change the way his firm caught fish in order to save dolphins. Children sometimes have more power than adults in these matters."

"Why would that be?" Katy asked.

"I think it's for two reasons," Ergon answered. **"First, kids speak from the heart; they often see problems and solutions clearer than adults. Another thing most adults realize is that kids will inherit the Earth; therefore, the future is really theirs."**

"So all these people, kids and adults, started using their natural resources and energy right." Mark was still trying to understand. "How did this buy time?"

Ergon explained that the pressure the people put on the national leaders by letter writing, meetings, hearings, publications, homemade videos of local environmental problems all began to have an effect on the nation's energy direction. Something that surprised many people was that more jobs became available. There was so much new work to do. New businesses sprang up to insulate homes and recycle and repair machines and goods. Companies found big markets for energy-efficient products like better light bulbs, recycled paper products, scrubbers to clean particles out of the air, better water heaters . . . the list was endless. As the countries were saving money by investing it in energy efficiency, they were creating money to help the poorer nations become developed in a sensible manner and not make the same mistakes they had.

"They bought time in two different ways," Ergon said. **"First, people agreed that fossil fuels were far too valuable to be burned up heating homes, making electricity, and powering automobiles. Those resources like coal, oil, and gas had to be used to make new products for the society. In other words, the society put its non-renewable resources in long-term energy usage, valuable products, rather than instant energy use. By deciding not to burn up the natural resources, they didn't draw on the reserves of the materials. This bought time to develop _renewable_ energy sources and deal**

with the other major planet-wide problem."

"Which was?" Mark asked anxiously.

"Too many people," Ergon replied. **"People cutting down tropical forests to live, people destroying the homes of animals who had been there for millions of years, people draining the planet of raw materials and water to conduct their daily lives. Just too many people drawing on the resources of the planet."**

"How in the world did they deal with that problem?" Katy asked. She knew it was a big one because she'd heard adults and kids she knew arguing about it.

"When people have the chance to be educated and earn enough money to have good things, they usually understand that having more and more children isn't the answer to a better world. The idea of large families goes back to ancient farming and religious practices that don't have much of a place in an overcrowded world. According to those beliefs, having lots of children gave a chance of survival to a tribe in case of a famine or other disaster. It made sense, in its time," Ergon said.

"But we don't live in tribes anymore, and neither did the people of Verdant. Did they, Ergon?" Mark asked.

"Actually, the people of this world came to realize that they were all one big tribe and that their survival and well-being depended on the health of the planet and the health of every living creature on it," Ergon answered.

While the group had been talking about the history of the planet Verdant, they had circled the beautiful globe about three times, passing over different parts on every circuit. They were still high above the atmosphere when Mark noticed a large object just above the horizon of the planet and approaching them rapidly.

"Ergon! What's that?" Mark shouted. By now the object could be seen as a large metal sphere headed directly toward them at a terrific speed.

"We've gotta move or that thing's gonna hit us," Mark cried. Ergon seemed to be laughing softly under his breath and made no attempt to change their direction. Before Mark could say another thing, the huge metal ball had passed right through them and disappeared behind. There wasn't a sound, but Mark got the impression of light and warmth during the instant the machine was passing through the group.

"Mark, I don't think you're ever going to learn that when traveling in energy form with me, you're not the same person," Ergon said.

"Was that a space station?" Mark asked in wonder.

"Yes," Ergon replied. **"We probably showed up as three blips on the radar screen and were gone before they noticed us."**

"We got run over by a space station. Wow!" Mark exclaimed, as if he couldn't believe any planet but Earth had space travel.

"Say, you two don't think the people of this planet turned their backs on science and progress when they turned away from energy waste, do you?" Ergon asked the children.

"Well, I hadn't really thought about it," Mark answered.

"I kind of thought they might all have to work on farms, or something," Katy added.

"With the planet's brainpower improving energy efficiency, the world had more money to spend on space science, health care, new and environmentally sound means of agricultural production—all kinds of things they didn't have before. Are you ready to see some of the methods they are using now?"

Both children said they were. Ergon took them into a deep dive down to the planet's surface. They were zooming in toward one of the larger landmasses. To Katy, the colors of the world didn't look much different from Earth at first. As they got nearer the surface, it seemed like there were more forests and less area given to cities and the sprawling areas that are often near them on Earth.

Ergon spoke. **"I will show you only technologies that are in current use on Earth and tell why they are in wider use here. I'm going to ask you questions, too, because now you should understand some of the energy reasons behind these decisions."**

Katy and Mark said that would be great.

They went first to a residential area. Neat homes could be seen in an attractive but irregular layout. There were lots of trees on the sides of the houses. None of them had leaves, and there was snow in the shadows, so it was winter here just like at home. The kids noticed that one side, the same side of

every house, was always left open and clear of trees. On the open side, the kids could see large panes of glass and darker colored walls.

Ergon explained that the clear glass-covered sides of the house were facing the south, or sunny side. **"This planet has one sun, similar in radiation output to yours. This part of the world has a cold and warm season, just like your hometown. So, what do you think the glass is doing on these houses?"**

"Good views?" Katy said, although she didn't seem very sure.

"Perhaps. But the real reason is solar heat," Ergon answered.

"I know about that. But it takes a bunch of machines and pumps and stuff, and it costs a lot of money! That's what dad told me," Mark said, proud of himself.

"Your dad's wrong, on this one. In some parts of your world, like Alaska and Siberia, solar heating is not practical because of extreme cold and limited sunshine. However, for most parts of this planet and yours, the proper design of a house can provide most of its heating and cooling needs. Can you imagine how much energy is saved? Good energy design doesn't have to cost more to build."

"But all that glass is going to make those houses cook like ovens in the summer," Mark responded, trying to win a point back from Ergon.

Ergon answered by telling Mark that if the glass was *not* placed right and

shaded from the summer sun, it could make the homes unlivable during the warm part of the year. But proper design would prevent that from happening. Ergon then pointed out the many trees on the sides of the homes. He told Katy and Mark that the trees do several good things for the house. They help block the winter wind from hitting the buildings, and they also shade the east and west sides and roofs of the homes in summer.

"Would a tree keep the big picture window in our living room cooler in the summertime?" Mark asked. "That window faces west and fries us in the afternoon."

Ergon told him that a shade tree planted on the west side of the window would save large amounts of energy in air-conditioning and make his house much more comfortable.

"I know what you're going to say next," Mark said with a laugh. "Another example of energy conservation making our lives better!"

"You're getting to know me pretty well, Mark."

Meanwhile, Katy had been looking intently at a group of houses that seemed to have more glass than the others. She asked Ergon, "What are those houses with more glass on the south? They look like greenhouses on the homes. And look, there seem to be plants growing all over the place inside, even though there's still snow on the ground!"

"Those are solar collectors, too. People call them solar green-

houses when they have lots of plants in them, sun spaces or sun rooms when the main purpose is solar collection or extra living space. A few people on Earth believe they are one of the most important discoveries of your century."

"Really?" Mark asked.

"Right up there with nuclear fission and the electric toothbrush," Ergon replied.

"And they were discovered in our century?" Katy inquired.

"Not really. The concept goes as far back as your ancient people. The principles behind the solar greenhouse were understood and greatly advanced in your century. On this planet, solar greenhouses have been in wide use for three hundred years."

Ergon explained that the concept of adding heat *while at the same time* supplying fresh-grown food throughout the winter was recognized on Verdant as having great importance. The devices had been put to many uses, depending on the climate and conditions. For instance, in areas with brackish or undrinkable water, greenhouses are used to distill pure water on the glass while they also grow food. In dry areas, the humidity the greenhouses provide to the home is welcome. At some locations on the planet Verdant, the building materials contain radon, a harmful gas, inside of them. To keep the buildings safe, solar greenhouses are used to bring in and heat outside fresh air. When fully planted, they become natural air purifiers. **"All in all, a simple device with a great many beneficial uses,"** Ergon concluded.

"Why don't we use them in more ways on Earth?" Mark asked Ergon.

"Right now, these passive solar technologies have little or no importance to your governments and leaders. But, a large and growing number of citizens are discovering the value. So, who knows? Maybe the leaders will catch up. If not, they'll be forced to take notice soon, anyway."

Ergon went on to explain that there is a big difference in using solar energy compared to using fossil fuels in buildings. Mainly, the local climate creates different building designs that work best in the area. A building can't be considered as a simple box with lighting, heating, and air-conditioning just stuck into it.

"On this world," he continued, **"we could go to a very hot climate and see the same solar principles used in a totally different way to cool the buildings naturally."**

But there was more to see here. Ergon took Katy and Mark down closer to the homes, and the children began to notice unusual details. Large parts of the roofing on the homes were covered with darkly covered areas. The material was very attractive. Sometimes it looked like shingles and in other places like expensive roofing tile. But the kids could tell it was neither of those things. Ergon was looking pleased. Katy spoke first.

"OK. We give up. You're looking very smug. So what are those things that make the roof so pretty?''

"The electric power generating station," Ergon said quietly.

"The whaaat?" Katy and Mark answered together, their mouths wide open.

"And the water heaters." It was then that Katy noticed that there were no ugly power lines going into any of the buildings. She also observed that no squat little propane or butane gas tanks sat alongside the houses.

"Let me explain what's going on here," Ergon started. Some of the space on the roof, he told them, was going to solar water heaters. Each house, by using water conserving devices and the solar heater, was providing the hot water needed by that family.

"They're not even ugly like the ones I've seen in town," Mark observed.

"I never thought those were ugly to begin with," Katy answered him.

"Some people don't know ugly when it hits them in the face," Mark snapped back.

"Well, the ones at home are no uglier than the power lines and TV satellites," Katy shot back, her voice getting louder.

"CHILL OUT!" Ergon barked. Both children were quiet immediately. **"You two haven't acted this way since I first saw you."**

Ergon continued his explanation. On Verdant, both the performance and the appearance of solar water heaters is more advanced than on Earth. Then he got into the electrical generating system.

"The power that comes from the roof is electrical generation by converting sunlight directly to electricity. The systems are called photovoltaic: 'photo' meaning light and 'voltaic' meaning electric. You have the same technology on Earth, but you haven't advanced nearly as far. On your planet, photovoltaic collectors convert to electricity under 10 percent of the sunlight striking their surface. Here, the collectors convert over 90 percent of the light into electricity. The power is stored in batteries that are far superior to yours."

"Well, why don't we do more to catch up to Verdant? Sunlight falls almost everywhere, everyday on Earth," Mark said with feeling.

"Sadly, it's a matter of commitment to old technologies and methods that cost billions of dollars and years of time," Ergon answered. **"Many scientists and government energy experts will tell you that this form of power can never be a big contributor to your**

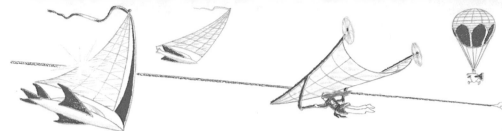

planet's needs. **But really, the technology isn't being heavily researched because the ramifications of this type of power are too threatening.**"

"What do you mean, ramifications?" Katy asked.

"The results, the changes that would happen if the renewable technologies were widely used," Ergon answered. **"You see, when you start greatly reducing the demand for conventional power, then all kinds of things start becoming possible. Look at this community."**

Ergon explained the energy links between all the renewable systems in the buildings below. First, because the homes are heated and cooled by the sun, no outside gas or oil is needed to do those jobs. The homes' hot water is the same, provided by solar energy.

"Each home is its own power generator," Ergon said.

For cooking, methane gas is provided by the organic wastes of the occupants, including their garbage. After all, Ergon explained, methane is what natural gas is mostly made of anyway, so why not get it directly from the organic matter in food and refuse. The small amount of extra heat the house or hot water might need in unusually cold or cloudy weather is also provided by the same home-generated methane gas.

The people on Verdant are growing much of their food right at home in solar greenhouses and gardens. By doing this, great electrical and mechanical energy is saved by not processing and moving that food around all over the country.

"All in all," he concluded, **"these homes are self-sufficient as far as energy is concerned, quite different from yours. If you think of a building as a power producer *instead* of a power consumer, then every family and business is a power station."**

As Ergon was talking, a small, sleek-looking vehicle pulled into the driveway of one of the houses below. Four humanoid beings got out of it and began walking toward other homes in the neighborhood. With very recognizable gestures, they waved good-bye to each other and made small talk Katy and Mark couldn't understand. The owner of the car went to the front end of the vehicle and pulled out a small cord that was then plugged into the house.

Ergon explained that the car, like the electricity in the house, is powered by the photovoltaic collectors on the roof. Earth has cars like this. Some had even gone across the entire continent of Australia, but most are limited in the distance they can go. The advances in photovoltaic collection and battery storage on Verdant allow the electric cars to go much farther and much faster than Earth models. If the drivers go on long trips or run low on power, there are "electric" service stations to quickly fill them up and get them on their way.

"All the transportation of Verdant is electric, so the greatest polluters of air—gasoline engines—don't exist."

"Let's look at some more of the planet, Ergon. I want to see what else they've done," Mark requested.

Ergon guided the group toward a large city center that looked to be about thirty miles away. They began flying in that direction, fairly low so that Katy and Mark could get a good look at the layout of the community. The biggest differences between here and home seemed to be more trees and greenery and less highways. They did pass over one large road that looked similar to the freeways on Earth. The children noticed that it was much less crowded and had fewer cars but more buses. Another difference was the clarity of the air.

Katy mentioned it to Ergon. "For a place that has as many people as this city, the air is too clear. Don't they make anything here?"

Ergon explained that the elimination of gas-powered vehicles had a lot to do with the clean air. But also, he went on to say, manufacturing processes had been improved on this planet. **"They don't create 'waste' when making products. Since all fossil fuels are extremely valuable here, every possible molecule is converted into something usable. Even the waste heat from the conversion of energy and materials is put to work by other companies and used in production. There are huge companies that make their money by separating and extracting the raw materials in containers, packaging materials, and worn-out products. Everything is recycled. Nothing is thrown away."**

"Do all the factories generate their own electrical power, like the homes?" Katy asked Ergon.

"Not all of them," he replied, **"Some of industry's production needs are too great. There is still an important place for centralized power generation, only they do that differently here, too."**

Ergon told Katy and Mark that all centralized power generating is based on renewable energy. There are many forms, some of which they'd seen with him on Earth, like hydroelectric plants and wind generators. Those are in wide use on Verdant. Like home design, power generating plants are built to take advantage of local conditions. For instance, large coastal cities like this one have links to ocean generating plants located offshore. These stations use the temperature difference between the water on the bottom of the ocean and the water at the top to drive generators. Other shoreline locations use the power of the tides, much like a dam on a lake, to create electricity from the mechanical power of moving water.

Desert and mountain locations often use hundreds of mirrors that focus the rays of the sun on a central boiler to make steam.

"And, of course, there are giant buildings in cities with lots of surface area to collect sunlight for photovoltaic power."

"Does it work just like the homes?" Mark wanted to know.

"The technology is the same, but the use is different," Ergon said. When the engineers and scientists of this world were devising all of the ways they could to replace fossil fuel electrical generation with renewable energy, they realized something important. Skyscrapers and all large buildings in cities are a resource. First, they have huge surface areas, perfect for photovoltaic collectors. The buildings are usually occupied during the day as offices and work spaces, so their power needs at night are small. Also, the tops of skyscrapers make excellent locations for wind power generators. The extra solar electricity generated during the day is put into storage to be used in other places. With superior design in business machines and lighting, the cooling needs of the buildings are far less than the same-sized structures at home.

"On this planet, all types of renewable electric power are fed into the grid. Everything from very small, stream-sized hydroelectric plants to photovoltaic skyscrapers are put into service. Much of the scientific effort has gone into better storage and transmission of electricity." Ergon concluded, **"On Verdant, they don't waste an erg of energy."**

"An erg? What's an erg? That sounds like a nickname for you," Mark said.

"Just a very small amount of energy, doing a job."

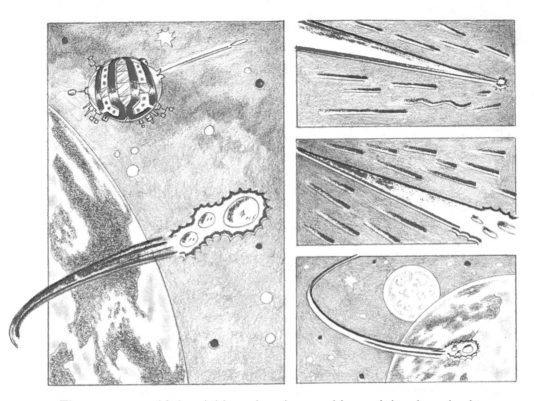

The creature told the children that they could spend days here looking at smart energy use in agriculture, machines, recycling, and transportation. **"Just think about this place and imagine what your planet could be: clean air, pure water, and room for all living creatures to grow and prosper."**

Now they were leaving the atmosphere of Verdant and heading up into the blackness of space. In the distance, Mark could see the space station that ran through them. From the edge of it, a large saucer-shaped rocket was heading out toward the stars.

"Is that a rocket? Where's it going?" he asked Ergon excitedly.

"The people of this planet are beginning to explore other worlds," Ergon answered.

Katy thought for a moment and said, "I guess that's only right. They've learned how to live on theirs."

Too soon for Katy and Mark, the return journey through the stars had ended. They were back home in their living room with Ergon. Both children

knew it was time for him to go. They tried to express their thanks.

"The way you'll thank me best is by putting the things you've learned with me to work," he said.

"Well, I'm going to go upstairs and make a list of projects we can do around this house to start saving energy right now," Mark stated.

"I've finally decided that my science project will be on ways to get energy out of trash," Katy said. "I'm also going to write our mayor a letter about the new recycling center that many people want. And I'm going to do it tonight."

In the background, the noise of a car engine could be heard.

"Your parents are home. I must leave," Ergon said. There was a touch of sadness in his voice. His body was shrinking into a smaller, more compact unit of energy as he prepared to travel.

"Where are you going next?" Mark managed to ask before the creature had entirely disappeared.

"There are quite a few politicians in Washington who really need a visit from me." The voice and Ergon were gone.

Glossary

Conservation   Finding ways to get the most out of the energy we must use. In some cases, like insulation in a building, conservation means keeping more heat energy in by stopping its flow through the skin of the structure. In other cases, like saving gasoline, it might mean car-pooling or simply driving less. Conservation of fossil fuels and natural resources is a lot like saving the last bite of a great candy bar until later during a long hike. You don't eat it now, because you know how much more you're going to enjoy it later.

Efficiency   A ratio of how much energy comes out of a system compared to how much energy went in. The number, if measured in percent, can never be greater than 100%. An electrical generator is one of the most efficient machines. Mechanical energy goes in, and 99% of the energy comes out in the form of electricity. A regular incandescent light bulb has electricity going in, but only 5% of the energy comes out in the form of light. The other 95% is waste heat.

   Because efficiency only measures the ratio of energy in to work out, and doesn't consider the source, it can be deceiving. For instance, an automobile gas engine converts 25% of the chemical energy of gasoline to mechanical movement. It is 25% efficient. A solar cell today would only convert 10% of the sunlight energy input into electricity. So it has a much lower efficiency. However, the solar cell causes no pollution, and its source of power is totally renewable. Which one is really doing a better job?

Energy   The ability to do work. It can be light, elecrical, chemical, kinetic, nuclear, magnetic, and a few others. Everything we do and everything that happens on earth demands the use of some form of energy.

Environment   Our perceived surroundings. Environment can be public or private. A private environment is usually controllable; we can surround ourselves with what we want. The public environment is shared by all and contains elements that can be measured and agreed on.

First Law of Thermodynamics   Energy flows from one place to another. It can never be created or destroyed. This law can be expressed in a lot of colloquial ways: "You get out of something what you put into it," "There's no such thing as a free lunch." This is the law that tells us that a perpetual motion machine will never be created, no matter how many times zany people try. If there's energy coming out of something, you can bet there's more going in. Look for the plug.

Fossil Fuels Incomplete products of the photosynthesis-respiration process from a long, long time ago. The most common and most used are coal, natural gas, and petroleum. The energy of ancient natural processes is locked into them and being burned at a rapid rate. Fossil fuels have only been seriously exploited in the last 100 years (less than 2% of humankind's written history), and most will be all gone in the next 50 years. Perhaps then we'll be the fossils.

Friction The resistance to motion of objects or surfaces rubbing against each other. The unwanted by-product is wasted heat. The efficiency of most mechanically driven machines (pumps, fans, motors, etc.) can be greatly improved, and the result will be conservation of fossil fuels and less wasted heat into the atmosphere.

Natural Resources Precious treasures that the Earth has provided. These can be minerals, chemicals, features of the land, or animals and humans.

Passive Solar Using the sun's position and local climatic conditions in architectural design to capture and distribute heat (or cooling) in a structure. In a well-designed passive solar building there is little or no reliance on mechanical systems such as fans or pumps. Some of the main elements of passive solar design are: direct gain, trombe walls, clerestories, roof ponds, convective loop collectors, and sunspaces (or solar greenhouses).

Photosynthesis The process by which plants convert sunlight into chemicals. Water and carbon dioxide are also needed. The output is oxygen and plant carbohydrates.

Photovoltaic Power The direct conversion of sunlight to electricity on the surface of a solar cell. This process offers great hope for the world as it has a minimum of polluting by-products and can bring electricity directly to the point of use. Photovoltaics are being increasingly used in remote locations and poor countries that can't afford huge centralized electrical power systems. They are also commonly seen in solid state electronic devices, like calculators and watches.

Pollution Contamination of the environment, especially by man-made waste. When the public environment or the earth itself is dirtied or worsened, it becomes everyone's problem.

Recycling Using the energy embodied in manufactured products over again. Environmental savings come from not having to process more raw material into new goods. Some materials like aluminum and glass lend themselves to recycling. Many plastics and industrial compounds in use today are not so easy to recycle.

Renewable (and Nonrenewable) Resources When a natural resource is renewable, it can be grown or used again and again. For instance, fuels made from corn or grain would never run out as long as there was good land for growing and harvesting. Sunlight, solar energy, is a renewable resource because it is available for use day after day. A nonrenewable resource, like oil, can only be used in its original form once. Each time we use a nonrenewable resource, it *should* cost more than the time before, because it can never be replaced and there is a limited amount.

Respiration The reverse of photosynthesis. In this process, plants and animals release carbon dioxide and other chemicals in order to live.

Second Law of Thermodynamics When left to itself in a natural system, heat will always flow from hot to cold. The sun must give the Earth energy, not vice versa. Your house on a cold January night will always give off heat energy to the atmosphere. If you know which of two objects you are observing is hotter, you know which way energy is flowing between them. The increased burning of fossil fuels is creating excess heat that must flow to the cooler atmospheric environment. We must find new ways to be more efficient, conserve, and use renewable technologies before the waste heat creates unsolvable problems for humans and other life-forms.

Waste A resource that someone has decided has no further use. In reality, wastes are packed with all kinds of energy that needs to be used.

Waste Heat A by-product of energy processes. Humans make waste heat when they excercise; it usually can't be recaptured. Industrial production and transportation make a great deal of waste heat that can be mostly recaptured.

REFERENCES

The organizations and publications listed below are dedicated to safe, renewable energy and a better environment.

Environmental Organizations

The Environmental Defense Fund, 1616 P Street NW, Suite 150, Washington, D.C. 20036
Global Tomorrow Coalition, 1325 G Street, Suite 915, Washington, D.C. 20005
Greenpeace, 1436 U Street NW, Washington, D.C. 20009
The Natural Resources Defense Council, 40 West 20th Street, New York, NY 10011
Worldwatch Institute, 1776 Massachusetts Avenue NW, Washington, D.C. 20036

Renewable Energy and Conservation Associations

These nonprofit groups advocating particular technologies are the best sources for the latest technical advances in their fields. They can also tell you if there are local chapters of their associations in your area. Most of these groups operate on a limited budget, so it is nice to include a self-addressed, stamped envelope when writing for information.

American Hydrogen Association, P.O. Box 15075, Phoenix, AZ 85060
American Solar Energy Society, 2400 Central Avenue, Suite B-1, Boulder, CO 80301
Geothermal Resources Council, P.O. Box 1350, Davis, CA 95617
Rocky Mountain Institute, 1739 Snowmass Creek Road, Snowmass, CO 81654
Solar Energy Industries Association, 777 N. Capitol St. NE, Suite 805, Washington, D.C. 20002

General Publications

Ecologue, edited by Bruce Anderson. New York: Prentice Hall, 1990.
   A wonderful composite of companies, products, and organizations dedicated to a greener earth and a sustainable energy future. A large section of it is aimed at kids' books and projects. Complete references give up-to-date information on a huge range of environmental organizations.
50 Simple Things Kids Can Do to Save the Earth, the Earthworks Group. Kansas City, Mo.: Andrews and McMeel, 1990. Great practical ideas for making a difference on Earth.
P-3, P.O. Box 52, Montgomery, VT 05470. Kids' environmental magazine.
PV Network News, 2303 Cedros Circle, Santa Fe, NM 87501. Newsletter on every alternate energy company and service you could ever imagine.

Renewable Energy Books

Some of these titles may be out of print, so check your local library.

The Food and Heat Producing Solar Greenhouse, Bill Yanda and Rick Fisher. Santa Fe, N.M.: John Muir Publications, 1981. The original book on designing and building greenhouses to heat homes and provide year-round vegetable gardens. The technical information is still pertinent and timely.

The New Solar Electric Home, Joel Davidson. P.O. Box 4126, Culver City, CA 90231. The classic how-to book on photovoltaics.

The Passive Solar Energy Book, Ed Mazria. Emmaus, Pa.: Rodale Press, 1979. All the design principles needed for passive solar buildings clearly explained.

Physics: Energy in the Environment, Alvin M. Saperstein. Boston: Little, Brown & Co., 1975. This is really a physics book, not a renewable energy book. The writing and illustrations are understandable, interesting, and entertaining. By teaching basic physics concepts, the author clearly demonstrates the need for developing renewable energy technologies.

The Solar Home Book, Bruce Anderson. Peterborough, N.H.: Total Environment Action, 1973. A good introduction to all types of practical solar use and building conservation techniques.

Sunspaces: New Vistas for Living and Growing. Pownal, Vt.: Garden Way Publishing-Story Communications, 1987. Beautiful photos and diagrams of homes throughout the United States which are heated with solar greenhouses.

Other Books from John Muir Publications

Adventure Vacations: From Trekking in New Guinea to Swimming in Siberia, Richard Bangs (65-76-9) 256 pp. $17.95

Asia Through the Back Door, 3rd ed., Rick Steves and John Gottberg (65-48-3) 326 pp. $15.95

Being a Father: Family, Work, and Self, Mothering Magazine (65-69-6) 176 pp. $12.95

Buddhist America: Centers, Retreats, Practices, Don Morreale (28-94-X) 400 pp. $12.95

Bus Touring: Charter Vacations, U.S.A., Stuart Warren with Douglas Bloch (28-95-8) 168 pp. $9.95

California Public Gardens: A Visitor's Guide, Eric Sigg (65-56-4) 304 pp. $16.95 (Available 3/91)

Catholic America: Self-Renewal Centers and Retreats, Patricia Christian-Meyer (65-20-3) 325 pp. $13.95

Complete Guide to Bed & Breakfasts, Inns & Guesthouses, Pamela Lanier (65-43-2) 520 pp. $15.95

Costa Rica: A Natural Destination, Ree Strange Sheck (65-51-3) 280 pp. $15.95

Elderhostels: The Students' Choice, Mildred Hyman (65-28-9) 224 pp. $12.95 (2nd ed. available 5/91 $15.95)

Environmental Vacations: Volunteer Projects to Save the Planet, Stephanie Ocko (65-78-5) 240 pp. $14.95

Europe 101: History & Art for the Traveler, 4th ed., Rick Steves and Gene Openshaw (65-79-3) 372 pp. $15.95

Europe Through the Back Door, 9th ed., Rick Steves (65-42-4) 432 pp. $16.95

Floating Vacations: River, Lake, and Ocean Adventures, Michael White (65-32-7) 256 pp. $17.95

Gypsying After 40: A Guide to Adventure and Self-Discovery, Bob Harris (28-71-0) 264 pp. $14.95

The Heart of Jerusalem, Arlynn Nellhaus (28-79-6) 336 pp. $12.95

Indian America: A Traveler's Companion, Eagle/Walking Turtle (65-29-7) 424 pp. $16.95 (2nd ed. available 7/91 $16.95)

Mona Winks: Self-Guided Tours of Europe's Top Museums, Rick Steves and Gene Openshaw (28-85-0) 456 pp. $14.95

Opera! The Guide to Western Europe's Great Houses, Karyl Lynn Zietz (65-81-5) 280 pp. $18.95 (Available 4/91)

Paintbrushes and Pistols: How the Taos Artists Sold the West, Sherry C. Taggett and Ted Schwarz (65-65-3) 280 pp. $17.95

The People's Guide to Mexico, 8th ed., Carl Franz (65-60-2) 608 pp. $17.95

The People's Guide to RV Camping in Mexico, Carl Franz with Steve Rogers (28-91-5) 320 pp. $13.95

Preconception: A Woman's Guide to Preparing for Pregnancy and Parenthood, Brenda E. Aikey-Keller (65-44-0) 232 pp. $14.95

Ranch Vacations: The Complete Guide to Guest and Resort, Fly-Fishing, and Cross-Country Skiing Ranches, Eugene Kilgore (65-30-0) 392 pp. $18.95 (2nd ed. available 5/91 $18.95)

Schooling at Home: Parents, Kids, and Learning, Mothering Magazine (65-52-1) 264 pp. $14.95

The Shopper's Guide to Art and Crafts in the Hawaiian Islands, Arnold Schuchter (65-61-0) 272 pp. $13.95

The Shopper's Guide to Mexico, Steve Rogers and Tina Rosa (28-90-7) 224 pp. $9.95

Ski Tech's Guide to Equipment, Skiwear, and Accessories, edited by Bill Tanler (65-45-9) 144 pp. $11.95

Ski Tech's Guide to Maintenance and Repair, edited by Bill Tanler (65-46-7) 160 pp. $11.95

Teens: A Fresh Look, Mothering Magazine (65-54-8) 240 pp. $14.95 (Available 3/91)

A Traveler's Guide to Asian Culture, Kevin Chambers (65-14-9) 224 pp. $13.95

Traveler's Guide to Healing Centers and Retreats in North America, Martine Rudee and Jonathan Blease (65-15-7) 240 pp. $11.95

Understanding Europeans, Stuart Miller (65-77-7) 272 pp. $14.95

Undiscovered Islands of the Caribbean, 2nd ed., Burl Willes (65-55-6) 232 pp. $14.95

Undiscovered Islands of the Mediterranean, Linda Lancione Moyer and Burl Willes (65-53-X) 232 pp. $14.95

A Viewer's Guide to Art: A Glossary of Gods, People, and Creatures, Marvin S. Shaw and Richard Warren (65-66-1) 152 pp. $10.95 (Available 3/91)

2 to 22 Days Series

These pocket-size itineraries (4½" x 8") are a refreshing departure from ordinary guidebooks. Each offers 22 flexible daily itineraries that can be used to get the most out of vacations of any length. Included are not only "must see" attractions but also little-known villages and hidden "jewels" as well as valuable general information.

22 Days Around the World, Roger Rapoport and Burl Willes (65-31-9) 200 pp. $9.95 (1992 ed. available 8/91 $11.95)

2 to 22 Days Around the Great Lakes, 1991 ed., Arnold Schuchter (65-62-9) 176 pp. $9.95

22 Days in Alaska, Pamela Lanier (28-68-0) 128 pp. $7.95

22 Days in the American Southwest, 2nd ed., Richard Harris (28-88-5) 176 pp. $9.95

22 Days in Asia, Roger Rapoport and Burl Willes (65-17-3) 136 pp. $7.95 (1992 ed. available 8/91 $9.95)

22 Days in Australia, 3rd ed., John Gottberg (65-40-8) 148 pp. $7.95 (1992 ed. available 8/91 $9.95)

22 Days in California, 2nd ed., Roger Rapoport (65-64-5) 176 pp. $9.95

22 Days in China, Gaylon Duke and Zenia Victor (28-72-9) 144 pp. $7.95

22 Days in Europe, 5th ed., Rick Steves (65-63-7) 192 pp. $9.95

22 Days in Florida, Richard Harris (65-27-0) 136 pp. $7.95 (1992 ed. available 8/91 $9.95)

22 Days in France, Rick Steves (65-07-6) 154 pp. $7.95 (1991 ed. available 4/91 $9.95)

22 Days in Germany, Austria & Switzerland, 3rd ed., Rick Steves (65-39-4) 136 pp. $7.95

22 Days in Great Britain, 3rd ed., Rick Steves (65-38-6) 144 pp. $7.95 (1991 ed. available 4/91 $9.95)

22 Days in Hawaii, 2nd ed., Arnold Schuchter (65-50-5) 144 pp. $7.95 (1992 ed. available 8/91 $9.95)

22 Days in India, Anurag Mathur (28-87-7) 136 pp. $7.95

22 Days in Japan, David Old (28-73-7) 136 pp. $7.95

22 Days in Mexico, 2nd ed., Steve Rogers and Tina Rosa (65-41-6) 128 pp. $7.95

22 Days in New England, Anne Wright (28-96-6) 128 pp. $7.95 (1991 ed. available 4/91 $9.95)

2 to 22 Days in New Zealand, 1991 ed., Arnold Schuchter (65-58-0) 176 pp. $9.95

22 Days in Norway, Sweden, & Denmark, Rick Steves (28-83-4) 136 pp. $7.95 (1991 ed. available 4/91 $9.95)

22 Days in the Pacific Northwest, Richard Harris (28-97-4) 136 pp. $7.95 (1991 ed. available 4/91 $9.95)

22 Days in the Rockies, Roger Rapoport (65-68-8) 176 pp. $9.95

22 Days in Spain & Portugal, 3rd ed., Rick Steves (65-06-8) 136 pp. $7.95

22 Days in Texas, Richard Harris (65-47-5) 176 pp. $9.95

22 Days in Thailand, Derk Richardson. (65-57-2) 176 pp. $9.95

22 Days in the West Indies, Cyndy & Sam Morreale (28-74-5)136 pp. $7.95

"Kidding Around" Travel Guides for Young Readers

Written for kids eight years of age and older. Generously illustrated in two colors with imaginative characters and images. An adventure to read and a treasure to keep.

Kidding Around Atlanta, Anne Pedersen (65-35-1) 64 pp. $9.95

Kidding Around Boston, Helen Byers (65-36-X) 64 pp. $9.95

Kidding Around Chicago, Lauren Davis (65-70-X) 64 pp. $9.95

Kidding Around the Hawaiian Islands, Sarah Lovett (65-37-8) 64 pp. $9.95

Kidding Around London, Sarah Lovett (65-24-6) 64 pp. $9.95

Kidding Around Los Angeles, Judy Cash (65-34-3) 64 pp. $9.95

Kidding Around the National Parks of the Southwest, Sarah Lovett 108 pp. $12.95

Kidding Around New York City, Sarah Lovett (65-33-5) 64 pp. $9.95

Kidding Around Paris, Rebecca Clay (65-82-3) 64 pp. $9.95 (Available 4/91)

Kidding Around Philadelphia, Rebecca Clay (65-71-8) 64 pp. $9.95

Kidding Around San Francisco, Rosemary Zibart (65-23-8) 64 pp. $9.95

Kidding Around Santa Fe, Susan York (65-99-8) 64 pp. $9.95 (Available 5/91)

Kidding Around Seattle, Rick Steves (65-84-X) 64 pp. $9.95 (Available 4/91)

Kidding Around Washington, D.C., Anne Pedersen (65-25-4) 64 pp. $9.95

Environmental Books for Young Readers

Written for kids eight years and older. Examines the environmental issues and opportunities that today's kids will face during their lives.

The Indian Way: Learning to Communicate with Mother Earth, Gary McLain (65-73-4) 114 pp. $9.95

The Kids' Environment Book: What's Awry and Why, Anne Pedersen (55-74-2) 192 pp. $13.95

No Vacancy: The Kids' Guide to Population and the Environment, Glenna Boyd (61-000-7) 64 pp. $9.95 (Available 8/91)

Rads, Ergs, and Cheeseburgers: The Kids' Guide to Energy and the Environment, Bill Yanda (65-75-0) 108 pp. $12.95

"Extremely Weird" Series for Young Readers

Written for kids eight years of age and older. Designed to help kids appreciate the world around them. Each book includes full-color photographs with detailed and entertaining descriptions of the "extremely weird" creatures.

Extremely Weird Bats, Sarah Lovett (61-008-2) 48 pp. $9.95 paper (Available 7/91)

Extremely Weird Frogs, Sarah Lovett (61-006-6) 48 pp. $9.95 paper (Available 6/91)

Extremely Weird Spiders, Sarah Lovett (61-007-4) 48 pp. $9.95 paper (Available 6/91)

Automotive Repair Manuals

How to Keep Your VW Alive, 14th ed., (65-80-7) 440 pp. $19.95

How to Keep Your Subaru Alive (65-11-4) 480 pp. $19.95

How to Keep Your Toyota Pickup Alive (28-81-3) 392 pp. $19.95

How to Keep Your Datsun/Nissan Alive (28-65-6) 544 pp. $19.95

Other Automotive Books

The Greaseless Guide to Car Care Confidence: Take the Terror Out of Talking to Your Mechanic, Mary Jackson (65-19-X) 224 pp. $14.95

Off-Road Emergency Repair & Survival, James Ristow (65-26-2) 160 pp. $9.95

Ordering Information

If you cannot find our books in your local bookstore, you can order directly from us. Please check the "Available" date above. If you send us money for a book not yet available, we will hold your money until we can ship you the book. Your books will be sent to you via UPS (for U.S. destinations). UPS will not deliver to a P.O. Box; please give us a street address. Include $2.75 for the first item ordered and $.50 for each additional item to cover shipping and handling costs. For airmail within the U.S., enclose $4.00. All foreign orders will be shipped surface rate; please enclose $3.00 for the first item and $1.00 for each additional item. Please inquire about foreign airmail rates.

Method of Payment

Your order may be paid by check, money order, or credit card. We cannot be responsible for cash sent through the mail. All payments must be made in U.S. dollars drawn on a U.S. bank. Canadian postal money orders in U.S. dollars are acceptable. For VISA, MasterCard, or American Express orders, include your card number, expiration date, and your signature, or call (800) 888-7504. Books ordered on American Express cards can be shipped only to the billing address of the cardholder. Sorry, no C.O.D.'s. Residents of sunny New Mexico, add 5.875% tax to the total.

Address all orders and inquiries to:
John Muir Publications
P.O. Box 613
Santa Fe, NM 87504
(505) 982-4078
(800) 888-7504